DATE DUE

MAR 3 1 2000	
OCT 0 6 2006	
JUL 2 7 2009	

Variable Speed Drive Fundamentals

Third Edition

Variable Speed Drive Fundamentals

Third Edition

Clarence A. Phipps

Published by
THE FAIRMONT PRESS, INC.
700 Indian Trail
Lilburn, GA 30047

Library of Congress Cataloging-in-Publication Data

Phipps, Clarence A., 1922-
 Variable speed drive fundamentals / by Clarence A. Phipps.--3rd. ed.
  p. cm.
  Includes index.
  ISBN 0-88173-310-5
  1. Variable speed drives. I. Title.
 TJ1051.P55   1999   621.8'5--dc21   98-48642
                                                            CIP

Variable speed drive fundamentals by Clarence A. Phipps, Third edition.
©1999 by The Fairmont Press, Inc. All rights reserved. No part of this publication may be reproduced or transmitted in any form or by any means, electronic or mechanical, including photocopy, recording, or any information storage and retrieval system, without permission in writing from the publisher.

Published by The Fairmont Press, Inc.
700 Indian Trail
Lilburn, GA 30047

Printed in the United States of America

10 9 8 7 6 5 4 3 2 1

ISBN 0-88173-310-5 FP

ISBN 0-13-021649-6 PH

While every effort is made to provide dependable information, the publisher, authors, and editors cannot be held responsible for any errors or omissions.

Distributed by Prentice Hall PTR
Prentice-Hall, Inc.
A Simon & Schuster Company
Upper Saddle River, NJ 07458

Prentice-Hall International (UK) Limited, London
Prentice-Hall of Australia Pty. Limited, Sydney
Prentice-Hall Canada Inc., Toronto
Prentice-Hall Hispanoamericana, S.A., Mexico
Prentice-Hall of India Private Limited, New Delhi
Prentice-Hall of Japan, Inc., Tokyo
Simon & Schuster Asia Pte. Ltd., Singapore
Editora Prentice-Hall do Brasil, Ltda., Rio de Janeiro

Table of Contents

	LIST OF ILLUSTRATIONS	ix
	FOREWORD	xiii
Chapter 1	INTRODUCTION	1
Chapter 2	VARIABLE SPEED DRIVE BASICS	3
	Mechanical Drives	4
	Hydraulic Drives	5
	Eddy Current Coupling	6
	Rotating DC Drives	7
	Solid State Systems	9
	Braking	10
	Definitions	11
Chapter 3	DC MOTORS AND DRIVES	15
	The DC Motor	15
	Speed Regulation	16
	Torque	17
	Control Systems	19
	Solid State Control	19
	Control Configurations	19
	Braking	22
	Ripple Factor	23
	Motor Maintenance Tips	25
Chapter 4	AC MOTORS AND DRIVES	31
	Induction Motors	31
	How Does the Motor Work?	34
	Current and Torque Relationships	35
	Control Methods for Induction Motors	36
	Full Voltage Starter	36
	Reduced Voltage Starters	38

	Solid State Motor Controller	40
	Smart Motor Controller	42
	Multi-speed Motors	44
	Synchronous Motors	45
	Synduction Motors	47
	Wound-rotor Motors	47
Chapter 5	MOTOR RESPONSE TO VARIABLE FREQUENCY	51
	Volts-per-Hertz	54
	Constant Volts Range .	55
	Characteristics of Variable Speed Drives	57
	Variable Voltage Input	60
	Current Source Inverter	61
	Pulse Width Modulation	63
	Scaler vs Vector Control	65
Chapter 6	HARMONICS	71
Chapter 7	PWM DRIVE PARAMETERS	77
	Starting the Drive	78
	Stopping the Drive	79
	Reversing the Drive	79
	Running Conditions	80
	Parameters	82
Chapter 8	MAINTENANCE OF PULSE WIDTH MODULATED DRIVES	93
	Components	93
	Generic Faults and Procedures	96
Chapter 9	BASICS OF APPLYING DRIVES	107
	Basic Relationships	107
	Replace DC Drive with AC System	111
Chapter 10	APPLICATION NOTES	117
	Define Machine Loads	117
	Peak Load	118
	AC or DC Drive?	118
	Define the Machine	118
Chapter 11	HEAT AND ENCLOSURES	127
	Defining Heat	128
	Temperature Calculations	128
	Heat Transfer Methods	129
	Examples	130

Chapter 12	ENERGY SAVING OPPORTUNITIES APPLYING MOTORS AND VARIABLE SPEED DRIVES	137
	Increasing Motor Efficiencies	137
	Other Energy Saving Opportunities	139
	Saving Energy with AC Inverters	143
	Payback	148
	Energy Auditing	150
	Useful Motor Calculations	153
	Applications	154
Chapter 13	CASE HISTORIES	161
	A Controlled Atmosphere Cold Storage Room	161
	Converting DC to AC Drives in a Pulp Washing Application	164
	A Utility-Subsidized Project	168
	Irrigation Controls	170
	Plating Facility Retrofit	171
Chapter 14	CONCLUSION	177
	Application: Breakaway, Acceleration, Peak Running Torques	178
APPENDIX A	Glossary	183
APPENDIX B	Formulas	191
APPENDIX C	Product Summary	199
APPENDIX D	Software	207
INDEX		209

List of Illustrations

Figure		Page
2-1	Mechanical Variable Speed Drive	4
2-2	Hydraulic Drive	5
2-3	Eddy Current Coupling	6
2-4	Rotating DC Drive	7
2-5	Solid State DC System	9
2-6	Solid State AC System	10
2-7	Torque	12
3-1	DC Speed/Torque Curve	18
3-2	EMI Grounding	20
3-3	Armature Voltage Regulator	21
3-4	Speed Regulator	22
3-5	Current Regulator	23
3-6	DC Bridge Configurations	24
3-7	DC Motor Cross-section	26
3-8	Simple Neutral Setter	27
4-1	AC Stator Circuit	32
4-2	AC Mechanical Structure	32
4-3	Three-phase Rotating Field	33
4-4	Rotor Current Generation	35
4-5	AC Current/Torque Relationship	36
4-6	NEMA Design Curves	37
4-7	Full Voltage	38
4-8	Reduced Voltage Starter	39
4-10	Solid State Reduced Voltage Starter with Isolation	40
4-11	Smart Motor Controller, defined	42
4-12	Smart Motor Controller Profile	43
4-13	Smart Motor Controller Profile	43
4-14	Smart Motor Controller Profile	44
4-15	Consequent-pole Motor	46
4-16	Synchronous Motor	47
4-17	Wound Rotor Motor	48

5-1	AC Current/Torque Relationship	52
5-2	Operating Range for Variable Speed Drives	52
5-3	Applying Variable Frequency to an AC Motor	53
5-4	Constant Volts-per-Hertz	54
5-5	AC Speed/Torque Curve	56
5-6	Connecting 3-Phase Winding to DC Source	57
5-7	Simulating 3-Phase from DC Source	58
5-8	3-Phase Motor Response to Square Wave Input	58
5-9	Variable Voltage Inverter (VVI)	60
5-10	VVI Wave Form	61
5-11	Current Source Inverter (CSI)	61
5-12	CSI Wave Form	62
5-13	Pulse Width Modulation (PWM)	63
5-14	PWM Wave Form	64
5-15	Vector Definition	68
5-16	Vector Representation	69
6-1	Third Harmonic Distortion	73
7-1	Photo, AC Drives, Courtesy of Allen Bradley Co. Inc, Milwaukee, Wisconsin	82
7-2	Parameters Affecting Volts-per-Hertz	87
8-1	Generic Components of PWM Drive	94
9-1	Torque	108
9-2	AC/DC Torque Comparison	114
10-1	AC vs DC Comparisons	119
10-2	EMI Grounding	124
10-3	Torque vs Voltage Drop	125
11-1	NEMA 12 Enclosure	130
11-2	Heat Exchanger	133
11-3	Air Conditioner	134
11-4	Vortex Cooler	135
12-1	Photo, Induction Motor, Courtesy of Allen-Bradley Co. Inc., Milwaukee, Wisconsin	138
12-2	DOE Required Motor Efficiencies October 1997	139
12-3	Power Factor	141
12-4	Efficiency Options	142
12-5	Affinity Laws	144
12-6	Fan Dynamics	145
12-7	Outlet Dampers, Power vs Flow	146
12-8	Inlet Vanes, Power vs Flow	146
12-9	Variable Speed, Power vs Flow	147
12-10	Energy Cost Comparisons	149

12-11 Performance Auditing Example..151
12-12 Controlled Atmosphere Storage...154
12-13 Variable Speed Refrigeration Pump 155
12-14 Roof Top Chiller 155
12-15 Dry Kiln 156
12-16 Vacuum Veneer Stacker 157
12-17 Water Treatment Plant 157
12-18 Pumping Plant...158
13-1 Controlled Atmosphere Storage...162
13-2 Torque, Square Rule Example 164
13-3 Pulp Washer Conversion..165
13-4 Replacing DC Drive with Equivalent AC Drive 167
13-5 AC vs DC Torque 167

Foreword

During a greater part of his fifty-year career in the electrical industry, the author applied, manufactured, and then commissioned many variable speed drive systems in USA and foreign countries. Since the products were usually custom-designed, there was a requirement to train the users in the art of operating and maintaining each application without overwhelming them with technical jargon.

There seems to be very few references that are written in *users'* language so they can understand the basics of applying their drives. The electrician does not have to know how to design a motor, or an electronic system—he needs to understand the operating principles of generic equipment engineered by others. The questions are,——"How does my motor work?"——"Why does it react to my load in this manner?"——"How do I improve its operation?"——"How do I maintain and trouble-shoot it?"

The material in this book has been prepared over a period of several years, and is basically a recap of the author's lecture series developed for user training, community college programs, and energy conservation seminars. In recent years, since retirement in 1990, the author has been involved in many 3-day "Drives" seminars in major cities across the USA. This book is an answer to his peers who suggested that the information should be captured in writing.

The intent is to give even an inexperienced technician or user a feel for the way different motors operate, and how they will respond to control systems. The material does not go into mathematical depths (calculus, etc.), but establishes basic relationships that can help one properly apply relatively simple drive systems with confidence.

Early chapters offer an in-depth review of different drive and control systems. Chapter 6, added in the 2nd edition of *Variable Speed Drive Fundamentals*, discusses the increasing concerns for harmonics. Newest technology, while trying to make AC motors run quieter, has introduced new harmonics problems that were not a concern a few years ago.

Chapter 9 discusses the basics of applying drives. Conversion from DC to AC drives is one of the important topics discussed in this chapter. A conversion case history is included in Chapter 13.

Chapter 10 deals with identifying machine characteristics, and installation precautions. Chapter 11 discusses the methods of controlling enclosure heat, and enclosure sizing.

Energy saving opportunities and useful information for calculations and estimates, applicable to utility energy conservation programs, are covered in Chapter 12, followed by case histories in Chapter 13.

Appendices C & D have been up-dated in this 3rd edition, with information, current in 1998.

If this generic discussion of drive systems helps make life easier for the electrician, application engineer, and user, its goal will have been met.

I continue to dedicate this book to my patient and loving wife, Betty, who kept the home fires burning during the many years that I was called upon to travel, and for her support in my retirement years as we prepared this text.

Clarence A. Phipps

Chapter 1

Introduction

For many years, the application of variable speed controls in industry was dictated by the process requirements, and limited by the available technology. Performance requirements took precedence over considerations of first cost, maintenance, and efficiency. The constant speed AC motor was usually used as the prime conversion from electrical to mechanical energy. To obtain variable speeds for machine operations, the motor required a secondary energy conversion using mechanical, hydraulic, or variable voltage DC generating equipment. While it was possible to develop very sophisticated control systems, every step in the conversion, including signal-amplifying exciters, added power losses, making the total installation relatively inefficient.

With the arrival of solid state semi-conductors in the 1960s, considerable headway was made to improve the efficiency of DC motor controls. Solid state AC speed controls were, for the most part, unreliable and very expensive. In the 1980s the development of large power transistors, with improved performance and reliability, started the AC drive revolution. This opened the door to applying the AC induction motor for direct variable speed control, without resorting to secondary electromechanical conversions. In many cases, the elimination of the extra hardware more than doubled the efficiency of the installation.

Since the energy crisis of the 1970s, it has become apparent that there is a finite amount of energy available. Because of the relatively low efficiencies that were tolerated in the past, the wasted energy was found to be contributing greatly to environmental pollution problems of the planet. The Environmental Protection Agency (EPA) was formed to provide some direction in correcting those problems.

The decision of the EPA to correct pollution problems has created a new industry with far-reaching impact. Power companies are required invest in expensive pollution control equipment, increasing their operating expenses and their charges for energy. During the past decade, industry and commerce has begun a "quiet revolution", upgrading facilities to take advantage of new energy conservation technology. The advent of reliable variable speed controls for the AC induction motor has contributed greatly to the success of conservation programs.

Many power companies now have subsidy programs available to industrial plants, and commercial users for energy saving installations. It is more profitable for them to recover energy from conservation than to build expensive new generation facilities.

In the Pacific Northwest, the original Washington State Energy Office, now a part of the University of Washington Extension Service, took the lead in promoting energy conservation programs, in cooperation with Bonneville Power Administration and local power companies. They have a continuing seminar program directed to Utility Energy Auditors, Building Operators, Hospitals, etc. Training resource people are recruited from an experienced pool of industrial experts in the fields of Power Quality, AC Drive technology, Lighting, HVAC, Refrigeration, and others.

The Electric Power Research Institute (EPRI) released a research report in the fall of 1992, called "Research and Development Plan for Advanced Motors and Drives". It outlines an aggressive strategy to shorten the lead time of advance drive and motor technology from concept to production. The stated goal is to save energy, improve power quality, and provide positive environmental impacts. One of the concerns that is expressed, is the "insufficient understanding of motor drive technologies, benefits, and proper application." The chapters that follow will address those concerns.

Equipment manufacturers have come a long way in designing reliable motor and drive equipment. Most of us in the electrical industry have applied motors to various constant speed applications where they operated at 60 Hz and provided performance that was reflected by the nameplate values. We now have to know how the motors perform over a wide range of frequencies (speed). For this reason, it is important for one to understand the proper *application* of these motors and drives, to prevent disastrous misapplications that can totally defeat the intended benefits.

Chapter 2

Variable Speed Drive Basics

For the purpose of the following discussions, a drive can be defined as a method of providing variable speed and/or torque to a machine input shaft. The most common source of power is the electric induction motor, which has a constant speed characteristic when connected directly to the power source.

Many processes require a change in speed during part of the operating cycle, or when a different production level is required. The weight or size of the materials may vary, as might be the case handling lumber products. For example, it is obvious that the optimum feed-rate through a saw would be slower for a large timber than for a small one.

Energy conservation may dictate a need to reduce power demand during periods of reduced or idle operating time. It is more economical to slow down a pump, for instance, than to continue to bypass fluids at full pressure.

With the **automated factory concept** that is predominate in industry today, control is essential to assure that different processes are synchronized to prevent waste, and coordinate the timing of related operations. This also includes the necessity for a means of communication between the drives and the command centers. As machines replace man in the repetitive, drudgery jobs, the coordinating decisions must be taken over by responsive drives and controls.

There are a number of different drive configurations that are common in industry. These include various mechanical, hydraulic, electromechanical systems as well as the DC and AC variable speed controls. We will briefly review some of the more common ones.

MECHANICAL DRIVES

One familiar system, is often referred to as a "Reeves" drive, named for one of the original manufacturers. It uses a variable pitch v-belt drive. In this case a pair of pulleys are used that have movable flanges. The flanges can be adjusted to different widths, so that a wide v-belt is forced to ride at higher or lower levels in the pulley, to provide different apparent pulley diameters. (Figure 2-1.)

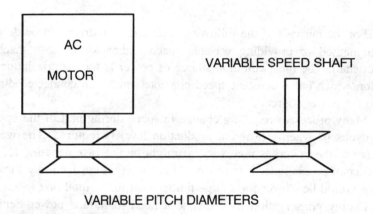

Figure 2-1 Mechanical Variable Speed Drive

The motor runs at a constant speed while the output shaft provides different adjustable speeds. The speed range is typically about 4 to 1. The drive also becomes a torque multiplier when the output shaft speed is slower than the motor shaft. This is a fairly inexpensive device, typically seen in lower power ranges, less than 50 hp.

Efficiency is relatively low, due to the belt slippage that is inherent to the drive. The pulleys tend to develop grooves when run for a time at a set speed. The speed changes then become "steps" rather than being infinitely variable. Speed settings may be manual or have motor-driven controls for remote control.

Maintenance of the belt, sheaves, bearings, and shifting mechanism can be become burdensome in a demanding application, but the drive works quite well in light to medium duty service. There are many of these being used for wood veneer dryer feed chains, for example. In this case, different speeds are selected to change the speed of the material through the dryer, as determined by thickness or moisture content.

HYDRAULIC DRIVES

A typical hydraulic drive includes a motor-driven pump, a hydraulic motor, reservoir, and control valve assembly, with associated piping and by-pass system. (See Figure 2-2.) Hydraulic drives give high torque and a wide speed range using a compact motor package. Control concept is relatively simple, as a valve controls the flow rate and direction of the fluid to the motor.

The pump runs at full speed at all times, so a pressure relief system is employed to by-pass fluid back to the reservoir during times of limited motor demand. Hydraulic fluids are non-compressible, so with positive displacement components, high performance can be attained.

Control options may be limited, and there are always the problems related to piping and fluids. There are environmental concerns from the leakage that frequently develops. Maintenance can be very difficult, as any foreign material in the hydraulic lines can quickly destroy a motor or pump due to the very close tolerances required to operate.

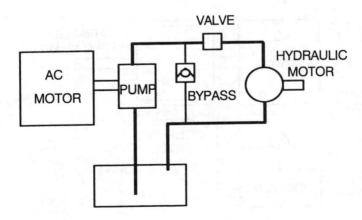

Figure 2-2 Hydraulic Drive

Efficiency of this configuration is relatively low, as a lot of heat is generated in normal operation, including the energy lost in the by-pass circuitry. First cost will likely be less than an all-electric drive, so operating expense and maintenance should be factored in when considering the application.

The more sophisticated hydraulic systems have variable stroke pumps that can control the flow rates without a by-pass system. These have much higher efficiencies and performance capabilities, but have similar maintenance requirements.

EDDY CURRENT COUPLING

The eddy current coupling is a device that provides an adjustable "slip" between the motor and the final drive. (Figure 2-3.) The motor operates continuously at rated speed, but the torque is delivered to the load at a reduced speed. The speed in this case is dependent upon the load.

The "stiffness" of the coupling is determined by the excitation level of a DC control field. High excitation currents will give an output speed near motor shaft speed, while reducing the current will increase the slip for a given load. The coupling should be considered a "torque" controller. The output speed will increase if the load is removed. By adding a tachometer feedback to the control circuitry, speed regulation can be improved. Efficiency is poor as there are always losses in proportion to slip. In fact, many of these devices use liquid cooling to dissipate the heat.

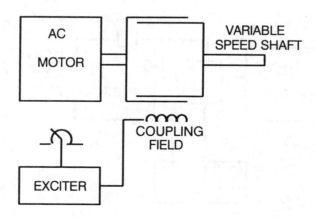

Figure 2-3 Eddy Current Coupling

Variable Speed Drive Basics

The principal maintenance item is the internal pilot bearing between the two rotors, which maintains the required air-gap. Some couplings have magnetic particles sealed in the gap to increase the field strength—but also the heat.

The horsepower range is limited. The system provides a very good "soft start" capability, and can absorb shock loads very well. Eddy current couplings are often used for punch press drives, veneer slicers, etc.

ROTATING DC DRIVES

There are many DC drives in service that are based on, what is commonly known as, the Ward-Leonard control concept. An AC/DC motor-generator set runs at a fixed speed. The DC generator armature is connected directly across the armature of a matching DC motor. Separate, controllable, field supplies are provided for the DC generator and DC motor to allow varying excitation levels for speed and torque control. (Figure 2-4.)

The motor field strength determines the available torque. The strength and polarity of the generator field determines the voltage and polarity of the armature circuit. Motor speed is proportional to the armature voltage, and its direction of rotation depends upon the polarity.

The field control systems usually employ some type of amplifier in the form of exciters, amplidynes, or solid state controls. This allows the use of very small control currents to control thousands of amperes in the armature loop circuit if necessary.

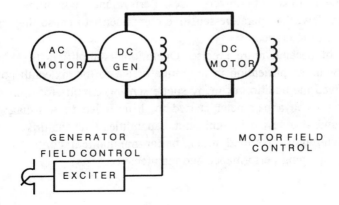

Figure 2-4 Rotating DC Drive

Performance can be very precise, with wide speed and torque ranges. There is a distinct advantage in that the system can absorb the energy from an over-hauling load, and return it to the power system. This is called "regeneration," and is a form of energy conservation.

The drive became very common after WWII, when many surplus DC machines were sold by the US Navy. There are hundreds of firms that assembled these systems, so there are also hundreds of control configurations—all accomplishing very similar functions.

Rotating DC drive machinery requires a lot of floor space. Environment may require total isolation in many cases, as the machines usually must be in open, ventilated spaces for cooling. Because of possible sparking, the installation cannot be in explosive atmospheres.

System efficiency is relatively low because of the number of rotating components, and heat producing controls and field windings.

SOLID STATE SYSTEMS

In the early 1960s, the development of the transistor, silicon controlled rectifiers (SCR) and operational amplifiers brought a whole new concept of control into being. It became possible to simulate the performance of a generator system, without using any rotating generators and/or exciters.

The efficiency of the system was greatly improved for several reasons. Friction and windage losses from the generator were eliminated, heat losses through the SCRs were much less than those generated by the generation equipment, and control circuitry operated at much lower power levels, thus with lower losses. Performance was improved, as there were fewer temperature-sensitive components to cause the system to drift out of tolerance.

Control became more precise. Operational Amplifiers (OPAMPS) could provide amplification in the range of ten to the sixteenth power! This allowed the addition of many subtle sensing circuits for immediate corrections when a parameter started to drift. It led to self-diagnostic circuitry and also new communication capabilities with the drive. All of these advantages provided much better mechanical control of the machines—helping performance and maintenance.

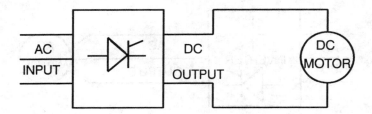

Figure 2-5 Solid State DC System

Solid State DC System

The solid state DC drive technology is very mature at this time. Almost any solid state DC drive can be expected to be reliable, and perform to the promised specifications. There are wide ranges of control options, controls are compact, have good efficiency, and are relatively inexpensive to procure. (Figure 2-5.)

The limitations are related to the complicated motor structure. The motor is an expensive item, with considerable maintenance required. The presence of brushes, with their occasional sparking, makes the drive difficult to apply in dusty or gaseous atmospheres, without a very expensive enclosure.

The SCR system causes considerable line disturbances, and also produces a varying *power factor (Pf)*. The varying SCR firing angle, with changing speed, changes the phase angle between the voltage and current sine waves. Power companies are very sensitive to Pf problems, and charge penalties for low Pf. (Ref Chapter 12). It is also difficult to provide a back-up operation for the DC system, if it becomes disabled.

Solid State AC System

If we consider the '60s to be the era of the DC drive, we might say that the '80s were the era of the AC drive development. The proliferation of AC induction motors in the industrial scene requires a reliable variable speed control similar to DC systems.

The control system is much more complex than the DC system. It is necessary to *convert* AC power to DC, and then *invert* it back to a variable frequency system compatible with the standard motors. (Figure 2-6.) This, while having all the advantages of solid state control, usually becomes more expensive to build than the DC package. (This price pre-

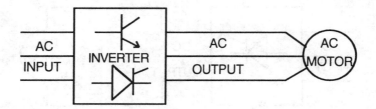

Figure 2-6 Solid State AC System

mium has been dropping very fast with improved electronics and sales competition.) The motor, however, has relatively low cost, and is often already in place. The motor has relatively high efficiency, is smaller than equivalent DC unit, and is an off-the-shelf product.

It is important to understand the differences between DC and AC systems when it comes to applications. They are *different*. Often the AC system has to be larger to provide the equivalent torque at certain operating speeds. When properly applied, performance is quite comparable. Another advantage with the AC system is that it can be by-passed if necessary, by connecting the motor directly to the line for a constant speed operation.

BRAKING

It is equally important to consider how a system is brought to rest, as well as how it is brought up to speed. Mass has inertia. We must add energy to get it in motion, and remove energy to stop it. Following are some methods that may be used for various applications.

Mechanical brake: This is typically an assembly using brake shoes, or discs which mechanically clamp the output shaft, similar to an automobile brake system. It typically is electrically released, and spring set. It may be AC, DC, hydraulic or air actuated. The energy is burned off by the friction heat.

Dynamic brake: This method uses the regenerative energy from the load, applied across resistors, to burn off the energy in the form of heat. Both AC and DC drives can use a form of dynamic braking.

Ramping: In this case, a controlled deceleration rate is used to slow the drive down. The rate must be slow enough so that the energy can be absorbed by the "system losses," without overheating or tripping off the control components. Ramping may not be able to handle heavy loads without fault.

DC Injection: DC current can be applied across two phases of an AC induction motor to bring it to a stop. The current must be limited, and the timing controlled. The heat in this instance is generated in the bars of the squirrel-cage rotor. If not properly controlled, the rotor will be damaged from repetitive braking. The heat may also cause deterioration of the stator windings.

Plugging: Not recommended. It is accomplished by throwing reverse power across the running motor. It can cause damage to equipment, motor, and controls, due the high instantaneous voltages, currents, and torques that appear.

Regeneration: Regeneration is the preferred braking method. It generally costs extra, as additional power components are required. In principle, the energy is returned to the power source with the least heat loss. The rate can be controlled by the system current-limit controls. Both AC and DC systems can be built with regenerative braking.

DEFINITIONS

Drive systems: As we discuss the drive systems, we are relating to the equipment between the power source, and the motor output shaft. This includes any isolation transformers (if required), the power conversion equipment, operator controls, and the motor.

Efficiency: The ratio of output power to input power. For example, output power divided by input power x 100 = % efficiency. The difference represents the system losses in the motor, control, and conversion equipment. Some typical examples follow.

DRIVE TYPE	100% SPEED	75% SPEED
M-G Set, DC	65%	61%
Eddy Current Coupling	86%	64%
Solid State DC	87%	85%
Solid State AC	86%	84%

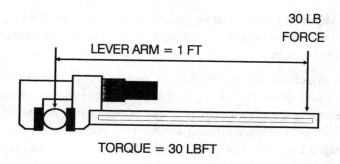

Figure 2-7 Torque

Torque: Torque is the turning force applied to a shaft, tending to cause rotation. It is typically measured in pound-feet, and is equal to the force applied (lbs), times the radius (ft) through which it acts. (Figure 2-7.)

Horsepower: HP is a measure of the amount of work performed over a period of time. To determine the HP, multiply the lb-ft torque times the RPM, and divide by the constant 5252.

$$HP = Torque \times RPM / 5252$$

Power factor(Pf): Pf is a measure of the time-phase difference between voltage and current in an AC circuit. It is the ratio of real power (kW) to the total apparent power (kVA). Pf is expressed as a percent, and is the cosine of the angle of phase displacement. Only when the voltage and amperage sine waves are in phase do we get pure power. (Ref. Chapter 12.)

Demand: Peak demand is another chargeable penalty from the power company. Demand determines how much the service equipment must be oversized to start the connected load. Peaks are metered over relatively short time intervals, and charges are made for the maximum load shown. Soft starts and drives are often touted to reduce the demand. While they do limit the in-rush currents, and thus line losses, there is very little *demand* advantage because total energy drawn to start the load is being metered. A drive will reduce the demand charges if the load does not have to be brought up to full speed each time it is started.

Heat: All electrical equipment generates heat due to the resistance losses, etc., of the components. Provisions must be made to dissipate this heat to prevent excess temperatures and thermal damage. It can be expressed in watts or BTU per hour (British Thermal Units). One watt = 3.414 BTU/hr. Enclosure sizing and cooling requirements are important factors to consider for successful installations. See the Chapter 11, entitled "Heat and enclosures," for a guide to handling heat problems.

Chapter 3

DC Motors and Drives

Direct current was the first successful form of electric power to be applied to motor control. Thomas Edison proved many of the concepts that are still in use today. While AC technology is now offering alternative drive controls for many applications, DC may still be the preferred system in many applications requiring very precise and high torque performance. (New vector-controlled AC systems now offer performance that rivals DC drives.) In the following material we will try to take some of the mystery out of the DC systems, with the intent to help mainstream electricians and operators understand the general characteristics of the DC motor and its control.

THE DC MOTOR

The DC motor is mechanically a bit complex, and often intimidates technicians who are assigned responsibility for its maintenance. There are two major parts to consider in this motor. One is the stator, or stationary "shell" which holds the field poles. The magnetic poles in small motors and servo units may consist of permanent magnets, but in most cases they will be wound with a "shunt winding." This winding is separately excited to allow precise control of the current. The magnetic flux density (torque) is related to the ampere-turns applied to the coil.

The rotating member is called the armature. It has many coils, wound in slots in the armature, and terminated to a current collector system, called the commutator. The commutator consists of many insulated copper segments which collect current from the brush assembly, mounted in the end-bell of the motor.

If the stator is wound in a four pole configuration, the armature will also have the same number of poles and brushes. The brushes may be single, or multiple, depending upon the amount of current that must be conducted through the armature. If the brush rigging has been set on its neutral position, there will be no difference in potential as it switches from bar to bar—and thus should not spark.

Since we can control the power separately in both parts of this motor, we can provide very precise speed and torque control. The field coils provide a fixed north-south relationship at all times, although the current strength may be varied to meet needs of the operation. Current is fed to the armature from a separate source, and because of the unique position of the coils at any given time, magnetic poles are produced.

These two magnetic systems interact to produce torque as the like poles repel, and unlike poles attract each other. As the armature turns, it continuously switches armature coils to maintain a constant magnetic relationship between the rotor and the stator.

SPEED REGULATION

We know that a magnetic field is produced around a conductor when current is passed through it. This is the premise of producing the magnetic fields, discussed above. It is also true that a voltage is produced in a conductor when it cuts a magnetic field. The voltage is dependent upon how fast the field is cut. This phenomenon is important to the speed control of the motor.

As the armature spins it generates a voltage which is always in opposition to the input source. This is called Counter-EMF(counter electromotive force), typically abbreviated as "Cemf." The motor will accelerate in speed until the Cemf nearly equals the input voltage. If the voltages were equal, there would be effectively no voltage difference across the armature to drive current through it. Armatures have very low resistance, so a difference of even a fraction of volt will pass considerable current.

If the motor slows down, due to an increased load, the Cemf will be less, so the apparent net voltage across the armature will increase, driving more current, and increasing the "ampere-turns" or torque. The additional torque will cause the motor to recover from the speed droop, and tend to stabilize the speed.

If armature voltage is reduced quickly, the Cemf will reverse the current in the system and provide counter-torque to bring the speed down to the selected value.

With a fixed field voltage, the speed of the motor will vary directly with the applied armature voltage and the direction of rotation will depend upon the polarity. If more speed is needed after we have reached maximum armature volts, we can use the Cemf theory, above, to reach the higher level. If we reduce the current in the field, the motor will have to run faster to generate the same Cemf to balance the input.

These relationships can be shown mathematically.

E_g = Cemf

S = Speed

Φ = Field Flux

K1 = Rpm per volt constant for the motor.

$$E_g = S \times \Phi \times K1$$

If we solve the formula for Speed, we have—

$$S = E_g/(\Phi \times K1)$$

Note that as "Φ" becomes smaller, "S" becomes larger. In fact if "Φ" became zero, "S" would become infinity. If you lose the field current, the motor may run away and destroy itself, as it tries to generate Cemf with only residual field flux.

Important—*A DC motor must always have its field excited before applying power to the armature circuit.*

TORQUE

Torque in a DC motor depends upon the summation of ampere-turns of the field and the armature windings. In any electrical motor, torque is generally synonymous with amperes. The pound-feet per armature ampere varies with changes in the field strength.

T = Torque

Ia = Armature current

Φ = Field flux

K2 = LbFt per ampere-turn constant for the motor.

$$T = I_a \times \Phi \times K2$$

As long as the field current is kept constant, the motor will have *constant torque* characteristics. It may have the same torque at 200 rpm as at 1000 rpm. When maximum armature voltage is reached, and field weakening is applied to further increase the speed, we are reducing torque in proportion to the increased speed. This is called the *constant horsepower* range of the motor.

DC motor nameplates often have a dual rpm rating on them, e.g., 1150/1750 rpm. This indicates that the constant torque range is from minimum to 1150 rpm, and that it is safe to extend the speed to 1750 rpm by field weakening for constant horsepower operation. In this example the 1150 rpm represents the *"base speed"* of the motor, when the field is at rated current, and the armature is at rated volts.

Figure 3-1 illustrates the torque and horsepower relationships for a 100 HP, 1750 rpm base speed motor operating into constant horsepower range to 200% of base speed.

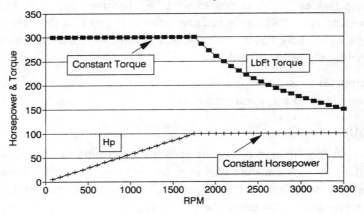

Figure 3-1 DC Speed/Torque Curve

CONTROL SYSTEMS

For many years, the typical control arrangement for DC drives used variable voltage generator systems to provide the discrete voltage and polarity requirements for the motor. Separate "exciters" (generators) are used to separately control the motor field current and the excitation for the generator field. The generator armature is connected directly to the motor armature, so it can follow the generator armature output, which determines the speed (voltage) and rotational direction (polarity) of the motor shaft. The motor field strength is controlled by the output of the separate motor field exciter.

Relatively small currents are required to control the exciters, which serve as amplifiers. A signal of a few milliamperes can control a generator putting out 1000 amps or more. There are many of these systems still in use that may be candidates for replacement by solid state control.

SOLID STATE CONTROL

The development of the *silicone controlled rectifier (SCR)* in the 1960s provided a new, much more efficient, power source for variable voltage DC drive systems. This eliminated the need for the large generators, and the maintenance and space that they required. System efficiencies were increased from about 65% to about 90%. This became a much more economical power source. The main drawback comes from the varying power factor (Pf) associated with the SCR, and the line disturbances (notching) that are reflected back to the power line. As a general rule, the DC SCR drive will always have an isolation transformer or line inductors ahead of it to help curtail these disturbances and limit the in-rush currents to the SCRs.

Considerable radiated EMI (electromagnetic interference) energy is also produced which can be very disturbing to sensitive computers and other electronic systems. In sensitive environments, a lot of attention must be paid to proper grounding and shielding, to protect low power instrumentation, etc. (See Figure 3-2.)

CONTROL CONFIGURATIONS

Figures 3-3, 3-4, and 3-5 illustrate the basic control schemes used with the solid state controls. The drive will always require a *reference* signal and a *feedback* signal which are of opposite polarity. The reference signal determines the desired output value and polarity, and the

SUGGESTED GROUNDING SYSTEM TO MINIMIZE E.M.I. NOISE

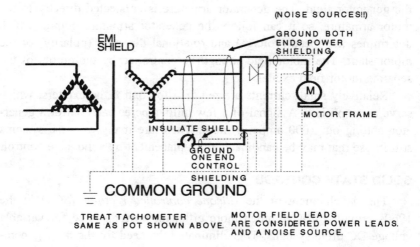

Figure 3-2 EMI Grounding

feedback is used to balance the control when the desired output is attained. This difference is called the *error* signal. (This is not defined as a mistake!) The source and accuracy of the feed-back determines how close the system "tracks" the reference.

The first example (Figure 3-3) illustrates the system used on most low-priced, simple drive systems. In this case, we see that a speed reference is introduced at a node ahead of a speed amplifier. We will consider this a positive signal for our example. By sensing the armature voltage at the output end of the drive, we can read the Cemf of the motor and use it as a negative feedback into the same control node as our input reference. When the error signal approaches zero, the system will stabilize.

Following the speed amplifier, you will see another node which is ahead of the current amplifier. Using a shunt or other device, the armature current is sensed and fed back on an inner loop to this node. The summation of the speed amplifier is positive, and the current signal is

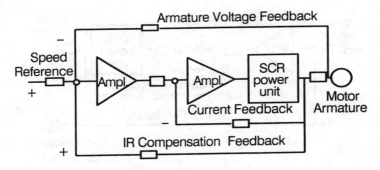

Figure 3-3 Armature Voltage Regulator

negative, so the current signal can be used to limit the current in the system. If the current is within an acceptable range, the speed signal is in control, but if current exceeds the norm, it will negate the speed control until the current is reduced to a safe value.

With this armature feedback system, the motor speed will tend to droop considerably between full-load and no-load, so the regulation (speed control) will be rather "soft." To help compensate for this droop, a feed-back, called an "IR Compensation" (IRComp), is added. This circuit senses the armature current, and feeds a small additive signal back to the speed amplifier node, meeting the reference and Cemf signals.

Now we have three signals meeting at the same node. The reference will be positive, the Cemf will be negative, and the IRComp will be positive. The IRComp adds to the speed reference to compensate for the droop created by the load. Since there are three signals meeting at a single input, we have a condition inviting instability. To set up this function on the drive, speed, Cemf, and current adjustments must be made with the IRComp turned OFF. Then with the motor running, a step change is made with the speed reference while observing the motor. The IRComp is gradually increased until the motor oscillates when the step change is made. It is then decreased to the point where the oscillation ceases. Under this condition, the regulation from no-load to full-load should be within 2% of full speed. The adjustment may have to be tuned a bit after the system warms up.

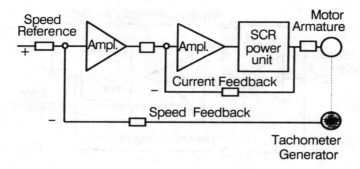

Figure 3-4 Speed Regulator

Figure 3-4 appears much the same as the first, except the Cemf and IRComp feedbacks are replaced by a *tachometer* feedback which reflects the exact motor rpm. Most sophisticated DC drives will use this method as accuracies in the area of 0.1% can be attained with a good tachometer. We are now looking at true speed, rather than an internal assumption from voltage and Cemf, for our feedback source. The quality of the tachometer determines the possible regulation accuracy of the system.

Figure 3-5 strips away all speed controls, and uses only the inner current loop for control. In this case, the reference determines the current (torque) needed, and the drive will operate at a speed that develops the desired torque. This mode is often used for winding and unwinding applications and many master/slave operations where motors must share torque. In reality, a maximum speed protection, or speed reference is used to prevent a run-away if the load is lost. Obviously, modern systems include many other unique features, depending upon the complexity of the application, but the above features are generally basic to drive systems.

BRAKING

Coast-to-stop oftentimes is not adequate to control the deceleration of a drive. Drives, as a rule, will have adjustments to control acceleration and deceleration rates when a step-change in reference is applied. Two other methods are used to stop a drive. These are *dynamic* braking and *regenerative* braking.

DC Motors and Drives

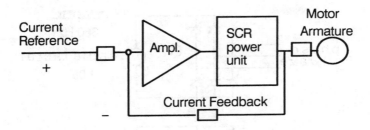

Figure 3-5 Current Regulator

Dynamic Braking

With dynamic braking, when a quick stop is required, the armature is disconnected from its source and reconnected across a resistor bank, while maintaining excitation on the shunt field. In this case, the Cemf (generator effect) of the motor is used to drive current into the resistor bank to absorb the energy. Counter-torque is created by the reverse current, and the effective braking is determined by the amount of current the resistors will allow to flow. The braking effect is strong at high armature speed, but falls off quickly as speed diminishes.

Regenerative Braking

Regenerative braking is a method of allowing the Cemf to regenerate directly back to the power line, with current limits controlling the braking force. The advantage of this method is that the energy is recovered, and not burned off as heat. It is also controllable, and effective to zero speed. This method requires the use of a dual SCR bridge as supplied with armature reversing systems. In some instances, a single armature bridge may be used in conjunction with a special reversing field supply.

A drive that is capable of both reversing and regeneration is called a "four quadrant" drive. This is an important feature when working with high-inertia, high-response reversing applications.

RIPPLE FACTOR

Ripple factor (sometimes called form factor) is a way of comparing SCR Drive power with that produced from a DC generator. Generator power is essentially pure DC, with practically no ripple. With a smooth DC current, there are minimum core losses in the motors, as the mag-

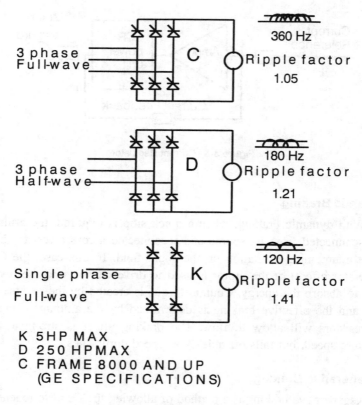

Figure 3-6 DC Bridge Configurations

netic materials are not subject to field oscillations. The ripple factor for a generator is considered as 1.0.

If we use a 3 phase, full-wave, 6 SCR power module, the ripple factor is 1.05. (Figure 3-6.) This means that on top of the DC power, there is a 360 Hz ripple that is equivalent to 5% of the rated power. This ripple will add some heat to the DC motor. In most cases this is not a problem.

Some manufacturers use a 3 phase, half-wave (3 SCR, 3-diode), power module. In this case, the ripple factor will be about 1.21. (Figure 3-6.) This means that there is a 21% ripple added to the DC which will create additional heat. The motor must be built to accept this wave form, using a laminated magnetic structure to minimize the core losses. Commutation is often very sensitive with these drives.

Single phase, full-wave SCR drives will have a ripple factor of about 1.41. (Figure 3-6.) The motors will have a considerably higher percent of losses. All modern DC motors, built for single phase sources, have laminated construction.

If motor size and economics allow, the 6 SCR three phase system is preferred. There are some drives that use unique transformers which, in effect, appear as six phase and further improve the ripple factor.

MOTOR MAINTENANCE TIPS

A system can have one of the best control systems available, and still not function properly. It is important that the motor has been set-up properly. New motors are usually properly adjusted at the factory, but after a period of operation, and possible repairs, the motor can have the original adjustments altered. This can reduce the efficiency of the motor, and perhaps create destructive sparking at the brushes. Consider the following set-up procedure to readjust the motor and minimize problems.

Adjust Brush Assemblies

Refer to Figure 3-7. Set the distance between the brushholder boxes and the commutator the same for all brushes. This can be done by slipping a thin wood or rubber shim under each box, and adjusting to same gap. The box should be set quite close to the commutator so that the brush works freely—usually with about 1/8 inch gap. If set too high, the brush will tend to drag and wear against the lower edges of the brushholder. The additional friction may cause the brush to chatter and spark if it cannot move freely.

Wrap a piece of paper around the commutator and drop the brushes in place. Mark the toe of every brush. (A strip of adding machine paper is very handy for this.) Now remove the paper and measure the distance between each toe mark. This distance must be equal. If not, then mark the paper with equidistant marks, slip it back on the commutator and adjust any errant brushholders, even if it requires a slight change in the gap.

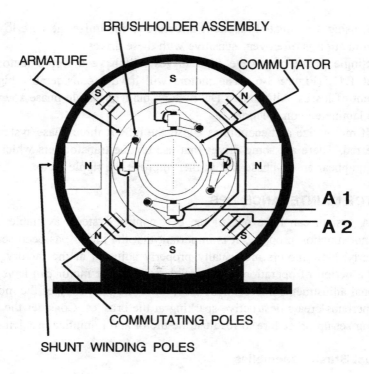

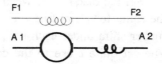

Figure 3-7 DC Motor Cross-section

Brush neutral setting

One of the most common reasons for poor commutation in a DC motor is an improper neutral setting for the brushes. This appears quite often after a motor has been disassembled or rewound. As a general rule, when disassembling a motor, the position of the brush assembly is marked so that it can be reset in the exact position when the motor is put back together. Sometimes, when rewinding an armature, the "throw" of the winding may miss the mark, terminating the coils one or two bars from the original position. The circuit in the armature is still correct, but

DC Motors and Drives 27

the brush pick-up point has been moved. Another source of trouble is that the brush holders may have a different angle of attack, thus pushing the toe or heel of the brush out of the neutral zone.

To set the neutral, we need a millivoltmeter. This can be the ammeter, disconnected from its shunt. We also need a battery source. Almost any battery, perhaps a 12v car battery can be used, with a means of switching on and off. Isolate the motor from any other controls and connect the millivoltmeter across A1-A2 armature leads. Connect the battery circuit across F1-F2 of the field circuit. (See Figure 3-8.)

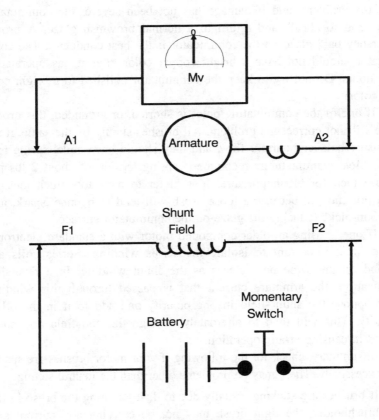

Figure 3-8 Simple Neutral Setter

While observing the meter, energize and de-energize the field circuit. If the meter kicks one way during the On cycle and reverses when Off, it indicates that the brushes are not on neutral. Loosen the brush assembly and rotate it a few degrees and repeat the test. If the needle remains stationary, you have found neutral. If, when you rotate the brush rigging, the meter kicks in opposite direction, you have passed the neutral point. When the needle stays at a "null" position, re-tighten the brush rigging. Now retest to be sure it did not shift during tightening. Repeat as necessary.

If you have a motor that is sparking, this simple test will often correct the problem, and if damage has not been severe, the commutator may tend to "heal" and return to a normal brownish glaze. A brown "Hershey bar" glaze on the commutator is the best conductor. The commutator should not have a bright copper color after it has operated a few hours. Do not try to keep the commutator polished to a bright copper color!

If one of the commutating poles is shorted, or grounded, this procedure will not correct the problem. All brushes should be the same grade to insure balanced current flow, and they should have equal spring tension. Most manufacturers recommend spring tension of about 2 lbs per square inch for brush pressure. It is better to have too much tension, than not enough, because a loose brush will tend to bounce, spark, and "carbon-pick", reducing the glaze on the commutator surface.

If one is using an older compound motor with a standard electronic drive, it is important to isolate the series winding. Series coils are wound on the same pole pieces as the shunt winding. In a reversing application, the armature current that is passed through this winding may oppose the shunt field in one polarity and add to it in the other polarity. This will tend to alternately demagnetize the field and compound it, causing erratic operation.

In summary, check for the following if your motor brushes are sparking excessively. There may be problems other than the neutral setting.

1. If brushes are sticking, usually due to dust jamming the brush in the brushholder, the tight brush may not be carrying any current and overloading other brushes. The culprit will usually be found on the underside of the commutator where it is hard to detect. "Snapping" all brushes periodically by pulling on their pigtails will detect stuck brushes, and also clean out the carbon dust.

2. If all brushes are not the same grade, soft ones will tend to "hog" current and not only wear faster, but sparking will damage the commutator. Always use the same grade of brush for all positions. If you have to use a different grade in an emergency, all brushes of the same polarity should have the same grade.
3. If you find a burned slot on the commutator, it is likely that the brushes are switching across a coil that has an "open" circuit. These open circuits often appear where a coil has thrown its solder at the connection to the commutator riser. These can often be repaired if too much damage has not occurred.
4. If you find a commutator bar that has lifted, it probably is a result of having stalled the motor too long in one position. Commutators usually require rebuilding or replacement from this damage, as the raised bar has not only been distorted, but also annealed so that its conductivity has changed.
5. Grounded or shorted interpole coils may cause sparking because they are no longer controlling the brush neutral point under changing loads.

While not all-inclusive, an understanding of the general concepts outlined above should give confidence to those who may be thrust into an unfamiliar DC environment.

Chapter 4

The AC Polyphase Induction Motor

All electric motors operate on the same electromagnetic principles. If you move a conductor through a magnetic field, a voltage will be generated in the conductor, proportionate to the number of lines of force that are cut per second. If the conductor has a closed circuit, current will flow. If the conductor is carrying current, a magnetic field will be induced around that conductor. Magnetic lines of force seek the shortest distance between poles, and interaction between the two magnetic fields produces repelling and attracting forces which develop **torque**. The strength and direction of the force is dependent upon the field strengths and relative polarities.

INDUCTION MOTORS

The most common industrial motor is the 3-phase "squirrel-cage" induction motor. By design, it has an electromagnetic stator which is wound with pairs of poles for each phase. The motor will always have three (3) circuits (or multiples of three) in a typical wye or delta configuration. This relationship is illustrated in Figure 4-1.

The rotor is an assembly of laminated magnetic material, with slots parallel with the shaft or slightly skewed from parallel. (Figure 4-2.) The slots hold conductors (usually cast aluminum), with shorting rings at each end of the rotor. The rotor fan is often cast as a part of the

3 - Phase AC Power

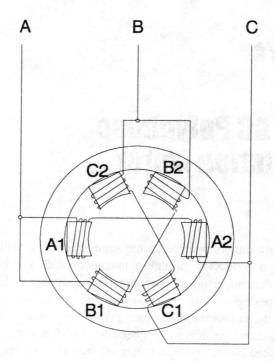

Figure 4-1 AC Stator Circuit

AC INDUCTION MOTOR

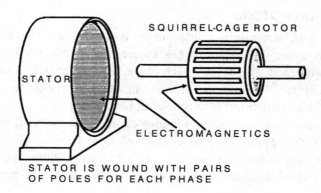

STATOR IS WOUND WITH PAIRS
OF POLES FOR EACH PHASE

Figure 4-2 AC Mechanical Structure

shorting ring. For all practical purposes, the rotor can be visualized as an assembly with many shorted coils.

The total assembly can be considered a transformer. The primary winding is in the stator, and the secondary is formed in the rotor. It is called an induction motor because all energy required in the rotor, for torque, is induced by this transformer action. The stator winding is arranged so that as the 3-phase, 60 Hz power is applied, the changing phase voltages will establish a rotating magnetic field. See Figure 4-3.

The synchronous speed of this field is dependent upon the number of poles and the frequency.

Synchronous Speed = Frequency x 60 / pairs of poles

Thus, a 4-pole machine will have a synchronous speed of 1800 rpm when operated on 60 Hz source.

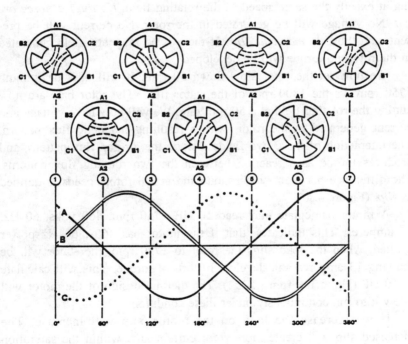

Figure 4-3 Three-phase Rotating Field

HOW DOES THE MOTOR WORK?

The following example considers a typical four-pole induction motor.

First, consider clamping the motor shaft so that it cannot turn, and applying 60 Hz power to the stator. The stator flux is now rotating at 1800 rpm. The magnetic lines of force are cutting the rotor bars at a maximum rate, so maximum voltage is generated in the squirrel cage "winding." Since the rotor is *shorted,* maximum current will flow in the bars. The primary of a transformer must supply all the energy required by the secondary, so we find that the motor is now drawing about 600% to 800% of rated current. If this condition is held more than a few seconds,—WE HAVE SMOKE!!! This is the locked rotor condition when a motor is first thrown across the line.

Now, let's look at the other extreme. Assume that we are able to spin the rotor to exactly 1800 rpm. In this case, the rotor bars are moving at exactly the same speed as the rotating field. *No lines of force are cut.* No voltage will be generated in the rotor. No current will be present, and thus, *no magnetic flux.* If there is no magnetic field generated in the rotor, no torque will be developed.

We now let the rotor slow down 50 rpm. It will now be running 1750 rpm, vs the 1800 rpm of the stator field. The rotor bars are now cutting the rotating field at a 50 rpm rate. We will now have voltage and current generated in the rotor, with a resulting magnetic flux pattern. The interaction of these two fields distorts the air-gap flux pattern, and produces torque as a reaction between the two rotating flux patterns. The difference between synchronous and actual rotor speeds is defined as *slip.* (Figure 4-4.)

A motor nameplate may read 10 HP, 1750 rpm, 460 volts, 60 Hz, 13 amperes. This indicates, that if the motor has 460v, 60 Hz power applied, when the load slips the rotor to 1750 rpm, the stator will be drawing 13 amperes and develop 30 lbft. of torque. This will calculate to 10 HP (Hp = T x rpm / 5252). The thermal design of the motor will allow it to run continuously under these conditions.

If we increase the load on the motor, slip will increase. The increased slip will create more rotor current and, within the saturation limits of the magnetic structure, more torque will be developed. If the slip is increased to about 75 rpm, we expect the current to rise to about 19 amperes and the torque to increase near 45 lbft. If this condition persists too long, winding insulation will be damaged by the excess heat.

The AC Polyphase Induction Motor

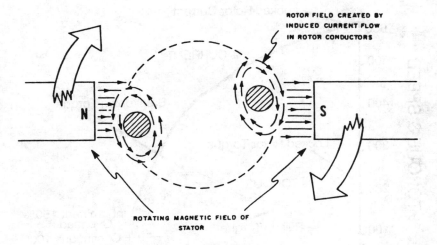

Figure 4-4 Rotor Current Generation

CURRENT AND TORQUE RELATIONSHIPS

When a motor is thrown across the line at 60 Hz, the instantaneous current will start at about 600% or more, and fall off on a relatively smooth curve to theoretically zero at synchronous speed. (Figure 4-5.) At the same time, a NEMA B design motor will have a starting torque of about 200%. Torque will fall to perhaps 175% *pull-up* torque at about 30% speed. It will then rise to a *breakdown* torque of about 250% at 80% speed (400% current), then fall off toward zero, meeting the current curve.

In most motor manufacturers' catalogs, you will find a chart similar to Figure 4-6, showing the speed/torque characteristics of different NEMA design motors when started across-the-line. The most common motor has a NEMA B characteristic described above. VSD manufacturers optimize their design around NEMA A and B motors. Each different motor design has a different *slip* characteristic. Since the slip reflects the relative current draw, one must compare actual motor amperage ratings with the nameplate of a VSD to be certain that it has adequate current capacity—we can't always depend upon the "Hp rating" of the drive to match all motors with the same apparent Hp rating.

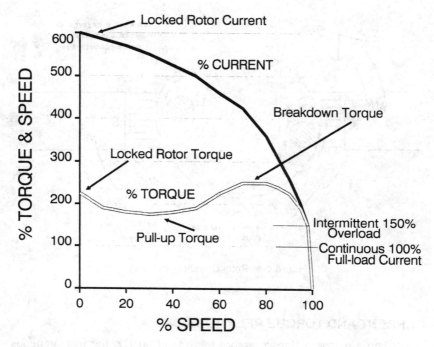

Figure 4-5 AC Current/Torque Relationship

CONTROL METHODS FOR INDUCTION MOTORS

We have been concentrating our introductory discussion on the common "squirrel-cage" induction motors, briefly mentioning the application of VFDs. It is important to understand the other most common methods of starting and controlling motors to put everything in perspective. As we discuss each controller, we will assume that necessary disconnects, protective circuit breakers or fuses, and overload relays are included.

Full voltage starter

The full voltage starter, often called *Across-the-line starter*, is the most common controlling device for the AC induction motor. (See Figure 4-7.) When the contacts are closed, full voltage is applied across the windings with the rotor at rest. This allows a high inrush of current in the area of 600% to 800% of the rated nameplate value. If the load cannot be accelerated quickly to full speed, the motor windings can be damaged by the excess heat, shortening the life of the motor. We have

The AC Polyphase Induction Motor

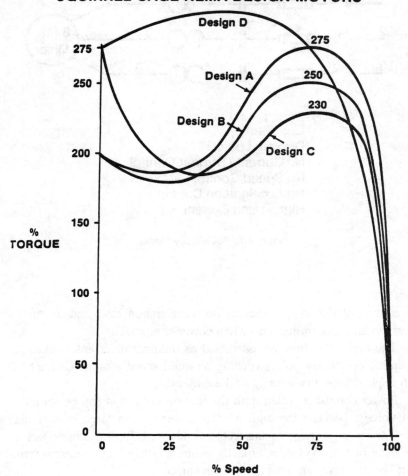

Figure 4-6   NEMA Design Curves

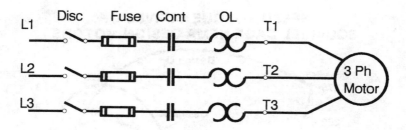

Line Start
Discrete Logic
No Current (Torque) Control
No Speed Control
No Acceleration Control
High Inrush Current

Figure 4-7 Full Voltage Starter

no control of the in-rush current nor acceleration rate, and the motor coasts to an uncontrolled stop when power is removed.

The contactors may be assembled as multiple units, interlocked for reversing control, or pole switching on multi-speed motors. Contact life will depend upon the severity of the duty cycle.

Power companies often limit the size of motor that can be started at full voltage because the high in-rush currents from large motors may draw down the voltage to unacceptable levels, effecting other users on the same line. If the feeder capacity is limited, they may require a "soft-start" control that can limit the in-rush currents.

Reduced voltage starters

Most non-electronic reduced voltage starters use auto-transformers to limit in-rush current and torque. Figure 4-8 shows the power circuit for "closed transition" auto-transformer starter. When a start is initiated, contacts 1S and 2S close immediately to connect the auto-transformers in a wye configuration, and the motor to the selected voltage taps. As the motor approaches running speed, contacts 1S open first, making the auto-transformers appear as series inductors, then contacts R close to

The AC Polyphase Induction Motor

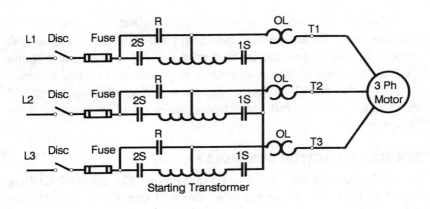

- Manual or Automatic
- Reduced current Inrush
- Reduced Starting Torque

- No Current (Torque) Control in Run Mode
- Acceleration dependent on Load
- No Speed Control

Figure 4-8 Reduced Voltage Starter

by-pass the auto-transformers—putting the motor across-the-line, and finally, 2S opens to remove the auto-transformers from the circuit.

Starting torque varies as the square of selected voltage tap value. An 80% tap allows about 64% starting torque, a 65% tap limits torque to about 42%. One should be aware of this limitation when applying these starters to a heavy starting load. If the torque is too low, the extended time it takes to accelerate the load may over-heat the motor and contribute to early insulation failure.

Manual reduced-voltage starters usually have the power switching contacts immersed in oil. One should never operate the starting lever while standing directly in front of the starter, because gases may build up from a sparking contact and literally blow the cover off. Manual operation should be done briskly to force quick contact closures and limit the arcing.

Automatic versions will use contactors to switch from *start* to *run*, and usually have adjustable timing relays that control the instant of transfer. Some use current relays that will not allow switching to run until the current has dropped below a pre-set value.

Whether using manual or automatic controllers, it is important to carefully determine the optimum switching point between too high in-rush current, and the motor's maximum thermal capacity. It is advisable to use the highest voltage tap that is acceptable, to provide best starting torque.

While the most common design uses auto-transformers, you will find some controllers that use a tapped resistor grid to accomplish the same functions.

SOLID STATE MOTOR CONTROLLER

Solid state motor controllers use Silicone Controlled Rectifiers (SCR) to ramp up the voltage at a controlled rate. One method brings the voltage up from 0 to 100% in a preset time. Another method maintains a preset maximum current level over time. The current control method provides higher starting torques, but may not protect the system from excess voltage drop as well. (See Figure 4-9.)

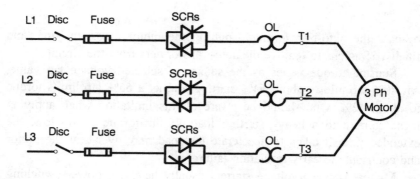

SELECTED ACCELERATION MODES

- Controlled Voltage Acceleration
- Constant Current Acceleration
- Current Ramp Acceleration
- Timed Linear Acceleration
- No Speed Control
- Limited Current (Torque) Control
- Optional Energy Saver Features

Figure 4-9 Solid State Reduced Voltage Starter

Some of these units have "energy saving" options that phase back the voltage when the load is light, to reduce the kW demand during idling time. This saving, alone, is usually not enough to justify selecting this type of controller, but probably should be activated if the unit is in place, and the duty cycle is applicable. If idling intervals are less than one minute, energy saving probably should not be applied.

One of the hazards of using this controller, is that when it is turned off, there is still standing voltage at the motor terminals. One should never work on the motor connections unless he knows that power has been isolated from the starter—and locked out.

Figure 4-10 illustrates a version of this starter that includes isolating and by-pass contactors. This is a much safer system because the motor **is isolated** from the power in the *off* condition. It also removes the SCRs from the circuit in the *run* mode which eliminates any line disturbances from the SCR, and extends the life of the solid state components because they are not *on* continuously. Energy saving mode is not available with this design.

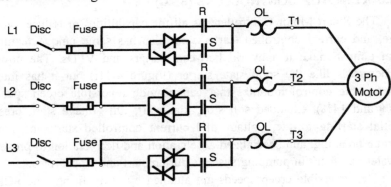

SELECTABLE ACCELERATION MODES

- Controlled Voltage Acceleration
- Constant Current Acceleration
- Current Ramp Acceleration
- Timed Linear Acceleration

- No Speed Control
- No Current (Torque) Control in Run Mode

Figure 4-10 Solid State Reduced Voltage Starter with Isolation

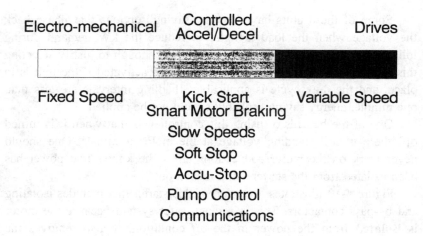

Figure 4-11 Smart Motor Controller, defined

SMART MOTOR CONTROLLERS (SMC)

The Smart Motor Controller has all the capabilities of reduced-voltage, and current controlled starters, but also has some unique features that almost make it a hybrid between starters and VFDs. The power structure is like the solid state starter (Figure 4-11.) but it has interchangeable control modules that can provide reversible creep speeds (7% and 14%), extended soft stop, kick-start, full voltage start, preset initial starting torque, voltage and current controlled starts, counter-torque breaking, and specialized acceleration and deceleration to prevent "water hammer" in pumping systems during start and stop cycles.

The reversible creep speeds are attained by "skip firing" the SCRs to produce an irregular rotating field in the motor. This is not used as a running mode, but only for *jog* duty with a controlled time and current adjustment. An application for this might be positioning a tumbler with the door on top for loading, then at the end of the cycle position the door at the bottom for unloading.

Kick-start might be useful to provide break-away torque to start a heavy load. It can also be applied with an initial minimum torque setting. Figure 4-12 shows an operating profile that includes creep speed, initial torque setting, kick start, and voltage ramp features.

Figure 4-13 illustrates a kick-start, voltage ramp, and soft stop features. Figure 4-14 illustrates slow-speed, voltage ramp, braking, and

The AC Polyphase Induction Motor

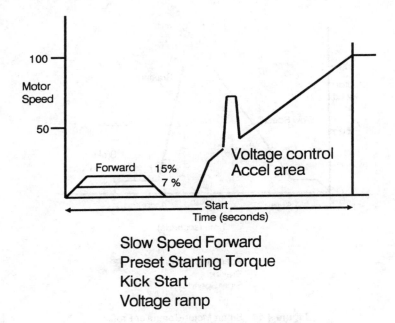

Figure 4-12 Smart Motor Controller Profile

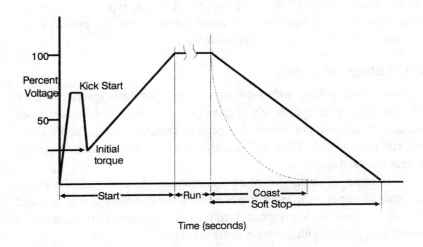

Figure 4-13 Smart Motor Controller Profile

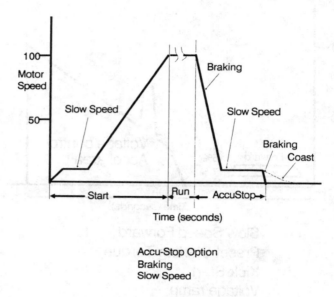

Figure 4-14 Smart Motor Controller Profile

slow speed positioning. Full voltage start might be used for emergency start in the event of a stalled, overloaded conveyor. Every combination has not been illustrated, but you can see that this type of control has many interesting applications, and could possibly provide essential performance without having to purchase a VFD.

MULTI-SPEED MOTORS

Two-speed motors, with separate windings for each speed, can have any combination of speeds, e.g. 1750/1150 rpm. They can be controlled like any other induction motor, but require separate, interlocked contactors for each speed. They will usually also require separate overload heaters for each speed.

One common design is called a *consequent pole* motor. The motor can have two speeds, using a single reconnectible winding, with a speed ratio of 2:1. Special interlocked 3 and 5 pole contactors are required to control this motor, with separate overload relays.

The AC Polyphase Induction Motor

When connected for high speed, the winding appears to be the same as any other induction motor in that a 1750 rpm speed motor will have 4 defined N and S poles.

When connected for 850 rpm, it would appear that all energized poles have the same polarity. These energized poles are called *salient poles*. Since every magnet must have both N and S poles, every salient pole will have a *consequent pole* of the opposite polarity. The motor effectively has eight poles and will run at half-speed (850 rpm).

The structure of the motor is somewhat compromised by this arrangement, and often the slow speed power factor will be considerably lower than at high speed, and efficiency will also be lower. Briefly, lagging power factor (pf) is a condition where magnetic loads tend to cause the current to rise out of phase with the voltage. Only in-phase current can produce torque. The out-of-phase current must be generated, and transmitted on the conductors, but does no work. Highest power factor is seen only when all voltage and current is in-phase. (See Glossary for definition of power factor.)

Figure 4-15 illustrates features of the consequent pole motor. If a VFD is applied to an existing motor, only the high-speed winding should be used. Some special duty motors have two (2) consequent pole windings, giving them a 4-speed capability. You must specifically order a motor for Variable Torque, Constant Torque, or Constant Horsepower applications.

SYNCHRONOUS MOTORS

A synchronous motor has a stator wound the same as any other induction motor, but its squirrel-cage rotor also has a DC winding that is terminated at a pair of slip-rings. See Figure 4-16. The motor starts as an induction motor, and when it reaches minimum slip, a sensing system is used to determine when passing poles are in line between the rotor and the stator. As the poles line-up, DC is applied to the rotor, and it pulls into step with the stator rotating field—running in synchronism. If the rotor is over-excited, it can be used to help correct a lagging power factor.

When a VFD is applied to the synchronous motor, the DC excitation is always applied, right from zero speed, eliminating the need for the pole sensing controls.

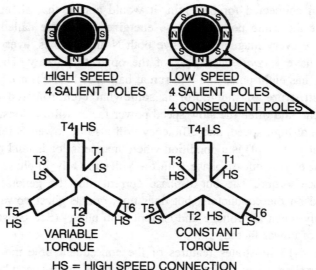

Figure 4-15 Consequent-pole Motor

The AC Polyphase Induction Motor

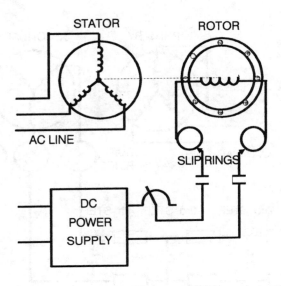

Figure 4-16 Synchronous Motor

SYNDUCTION MOTOR

The synduction motor is a special purpose motor that has a squirrel-cage construction, but the rotor has lobes that tend to form discrete poles. When this motor reaches minimum slip, if not overloaded, it will snap into a synchronous speed. Its torque will be determined by the angular displacement between the rotor and stator rotating fields. If excess load is applied, it will "slip poles", changing between induction and synchronous modes. These are found most frequently on synchronized conveyor lines, operating from a common variable frequency source. If the motors slip poles, the cogging action can produce high current excursions that may take a VFD off line. They cannot be operated successfully if subject to high load surges.

WOUND-ROTOR MOTOR

The wound-rotor motor is a very interesting induction motor, with very special capabilities. It is truly an induction motor, but has a 3-phase winding on the rotor, terminated at 3 slip rings, instead of squirrel cage construction.

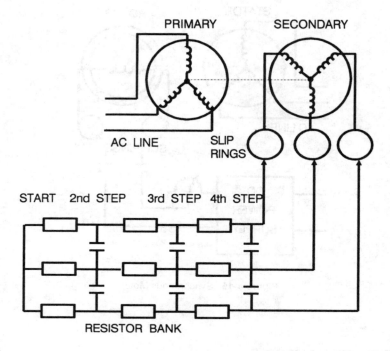

Figure 4-17 Wound Rotor Motor

The rotor windings have no connection to the power source. The rotor in this case becomes the secondary of a transformer. All current in the rotor is derived from the primary, and is dependent upon the slip of the motor, and the amount of resistance that is in the rotor circuit. See Figure 4-17.

The total energy passing through a transformer will vary according to the demand on the secondary winding. If we have high resistance in the secondary circuit, the current in the transformer will be reduced. In this case, limiting the current limits the torque the motor can produce. Note that as the current in the rotor is allowed to increase, the primary current also increases, compounding the available torque, as more magnetic flux is added to both the primary and the secondary fields.

The nameplate of the wound-rotor motor will have the usual values of Hp, Rpm, Volts, Amps, etc.. It will also have a secondary voltage rating which is the voltage that will appear on the slip rings when the rotor

The AC Polyphase Induction Motor

is at rest, and line voltage is applied to the stator. There will also be a current rating for the rotor that identifies rated current capacity of the rotor windings. We use these values to design the torque characteristic that we want the motor to have.

For instance, we will assume that the motor has a secondary voltage rating of 360 volts, and the winding is capable of 65 amperes. If we can put enough resistance in this circuit to limit the current to 65 amperes at stall, the motor will have 100% torque and draw 100% line current when starting. We would not see the 600-800% starting current of the squirrel-cage motor.

If the circuit is similar to Figure 4-17, assuming 360 volts at the slip rings, there would be 208 volts on each leg of the wye-connected resistor bank.

```
360 volts / 1.73 = 208 volts
208 volts / 65 amps = 3.2 ohms per leg
```

So if we can install 3.2 ohms in each of the slip ring circuits, and connect them in a wye configuration, this motor will have 100% torque available when full voltage is applied to the stator. The wattage of the resistors would have to be about 13 kW if this condition were continuous, but if only applied for a few seconds, would probably only require a 1300 watt short time rating.

The above condition is at 100% slip. As the motor accelerates, the relative slip will be reduced, so less secondary voltage is available, and we can reduce the resistance to maintain approximately constant current in the secondary circuit. The second step might reduce the resistance to 1.6 ohms, the third step to 0.8 ohms, the fourth step would short-circuit the slip rings. From this point on the motor operates like a squirrel cage induction motor.

It can be seen from the above that we can have almost infinite control of the motor torque (amperes) by varying the secondary resistance. This is **not** a speed control system. If the load is removed at any grid step, the motor will accelerate to its speed at minimum slip.

When converting a wound-rotor system to VFD control, you can usually short out the slip rings, and connect the VFD to the stator. Some wound-rotor motors will not start without some minimum resistance in the rotor circuit, so one would have to test the motor first to see if it will start with shorted slip rings. As an example, if you were converting a crane travel or bridge control to VFD to control the swing of the load better, you might have to leave the minimum resistance step in the circuit to guarantee motor starting.

It is interesting to note that a wound-rotor motor can be driven by a variable speed DC motor, and take power off the slip rings as a frequency changer.

If multiple wound-rotor motors are used to drive synchronized conveyors, the rotors can all be connected in parallel. With a common variable frequency applied to the stators, all motors will lock-step and maintain constant angular rotor relationships. In other words you have a "synchro" drive. One application of this can be found in some sheetrock manufacturing plants where any difference in conveyor speeds would rupture the panels.

The wound-rotor motor is more expensive than the common induction motor, and has fallen out of favor in many applications because few people have any idea how they really work or how they can be used to an advantage.

Chapter 5

Motor Response to Variable Frequency

In our discussions up to this point, we have considered the ways of controlling various induction motors, operating on 60 Hz. What happens when I try to operate at other frequencies?

Earlier, we briefly discussed the starting characteristics of the squirrel-cage induction motor. In the area of about 150% torque or less, the current and torque curves share the same line. See Figure 5-1. In other words, if we can look at the current in this region, we can assume that the torque and slip are linear with the amperage. All variable frequency drives operate in this linear area and use the current as a reflection of motor performance. See Figure 5-2.

If our VFD can provide the conditions suitable to keep the AC motor operating within this "linear" range, the motor should respond properly. In this case we are controlling the motor indirectly, contrasting with DC systems where we force performance by separate field and armature control.

VFDs have nominal Hp ratings, but one should always compare the *ampere* rating with that of the motor that will be used. Vertical pump motors often have higher current ratings than equivalent horizontal motors and may require a larger drive. Constant torque drives usually have an overload rating of 150% of their *nameplate* amperes. Note that this overload value may be higher than the motor rating, and may have to be adjusted accordingly.

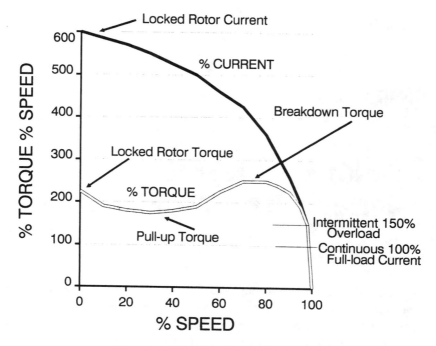

Figure 5-1 AC Current/Torque Relationship

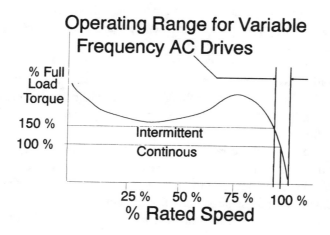

Figure 5-2 Operating Range for Variable Speed Drives

Motor Response to Variable Frequency

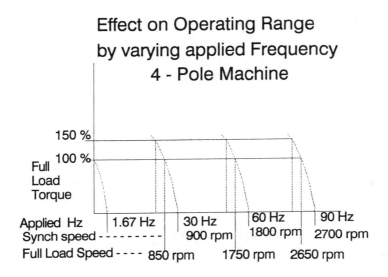

Figure 5-3 Applying Variable Frequency to an AC Motor

Looking at Figure 5-3, you will see a part of the torque curve, taken from the previous figure, showing that at 60 Hz, the subject motor has a synchronous speed of 1800 rpm, and the shaft speed at 100% torque is 1750 rpm—this is 50 rpm slip. In this case, whenever the slip is 50 rpm, the motor is developing 100% torque.

At 30 Hz, the new synchronous speed is 900 rpm. To get 100% torque the motor must slip 50 rpm, so the new shaft speed is now 850 rpm. Note that synchronous speed and frequency have a direct relationship, but the shaft speeds do not. Cutting the frequency in half produced one-half synchronous speed, but the shaft speeds were off-set by the rated slip. *Whenever calculating speed and frequency relationships, always use synchronous speeds.*

Examples: (Use motor discussed, above)

1. Find required frequency required for 700 rpm on the shaft.

    700 rpm shaft + 50 rpm slip = 750 rpm synchronous speed.
    (750 rpm synch / 1800 rpm rated synch) x 60 Hz = 25 Hz

2. Find shaft speed at 70 Hz, with rated slip.

    (70 Hz / 60 Hz) x 1800 rpm synch = 2100 rpm synch
    2100 rpm synch - 50 rpm slip = 2050 rpm shaft speed

It is interesting to note that 1.67 Hz is equal to 50 rpm slip. Disregarding other losses, the application of 1.67 Hz can theoretically produce 100% torque at zero speed. This illustrates the fact that we can apply increasing frequency (slip) until there is enough torque to start our load, and never see the 600-800% in-rush currents. By the time we reach 2 or 3 Hz, we will have produced over 150% torque which should start any load that is matched to the motor.

VOLTS-PER-HERTZ

When working with variable frequencies we find that *inductive reactance will vary directly with the frequency* [X_L (ohms) = $2\pi fL$]. This means that as frequency increases, the resistance to current flow (ohms) increases at the same rate, and vice versa. If we are to maintain constant current capabilities, we have to increase voltage in proportion to the increasing frequency.

A 460 volt motor has a 7.6 volts-per-hertz ratio. (460 / 60 = 7.6 volts-per-hertz.) A 230 volt motor has a 3.8 volt-per-hertz ratio. (230 / 60 = 3.8 volts-per-hertz) See Figure 5-4. At 30 Hz our 460 volt motor only requires 230 volts to supply rated current (torque) at rated slip. *As long as we can maintain a constant volts-per-hertz ratio, the motor will have constant torque characteristics.*

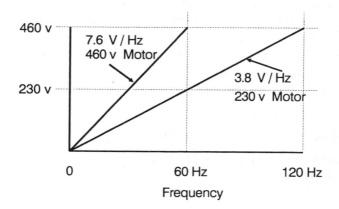

Figure 5-4 Constant Volts-per-Hertz

The nameplate values establish our base speed and voltage for the AC motor. It will have constant torque characteristics when operated from zero to base speed if the volts-per-hertz ratio is maintained.

Figure 5-4 also shows a 230 volt motor operating at constant volts-per-hertz from zero to 460 volts, and 120 Hz. In this case the 460 volt drive has been set-up for 3.8 volts-per-hertz. When the 230 volt motor reaches 60 Hz, it is developing rated horsepower. As frequency and voltage are increased above 60 Hz, the motor still has constant current capabilities, and by the time it reaches 120 Hz, it is developing double its rated horsepower. The insulation of the motor is not in jeopardy since it is probably rated for 600 volts or more, if it is a nine-lead, dual-voltage motor. The current has not exceeded its rating because of the increasing inductive reactance. Motor cooling is increased by fan speed. Our only concern is for the mechanical limits of balance and bearings. If this were a 5 Hp motor, the drive would have to be rated for 10 Hp to carry the higher current required for a motor connected for 230 volt operation.

An application for this might be a swing-saw, where it is important to keep the swinging mass to a minimum. As an example, the saw might require 10 Hp @ 2200 rpm. A 5 Hp, 1150 rpm motor might be used and operated to about 113 Hz to do the job.

CONSTANT VOLTS RANGE

In most installations, when the motor has reached its base speed, it has also reached maximum available voltage. If we want to extend the speed beyond this point, we can increase the frequency, but we cannot maintain the volts-per-hertz ratio for lack of voltage. Our inductive reactance is still increasing with frequency, but we have no way of compensating for it. At rated slip, we can no longer produce rated torque (current).

With reduced current, we have less flux in the stator and the rotor, (since all rotor energy must be derived through the stator) so our torque falls off as a square, and we are starved for voltage. The percent of available torque at rated slip, at any point beyond base speed can be calculated by dividing 60 Hz^2 by the new $frequency^2$, e.g. $(60 / f)^2$. At 90 Hz the motor would have 44% of rated torque at rated slip.

Many manufacturers call this speed range the "constant horsepower" range similar to the characteristics of a DC motor. In reality, true constant horsepower operation is not available at rated slip. Speed regulation will fall off, as slip must increase. A DC motor running at 150%

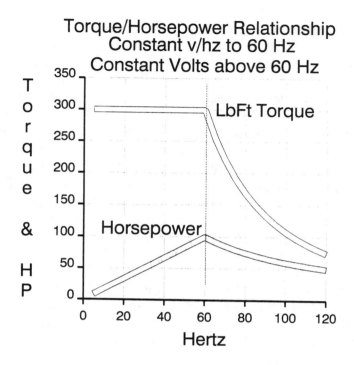

Figure 5-5 AC Speed/Torque Curve

speed would have 67% torque. The AC motor would have to slip 150% to equal that torque. The AC motor can approximate constant horsepower up to about 90 Hz, but beyond that the excess slip will likely pull the motor down to the break-down torque for that frequency.

Figure 5-5 shows a torque and horsepower graph for a 100 Hp AC motor. You will see that at 200% speed, the torque is down to 75 lbft, and it is only producing 50 Hp at rated slip. A DC motor under the same conditions would produce 150 lbft of torque and 100 Hp. From this you can see that one must be very careful when replacing DC with AC drives, especially when operating at extended speeds.

The concepts discussed here our very important to understanding the basic relationships between motors and drives.

- Torque is related to slip
- Torque is related to volts-per-hertz ratio
- Frequency is related to synchronous speeds

- Torque falls off as a square of the frequency in constant voltage range
- Note the mathematical relationships and calculations.

CHARACTERISTICS OF VFDs

Prior to the 1980s, the usual way of providing variable, or adjustable frequency for control of AC motors was to use an alternator, driven by a DC motor. DC excitation to the slip-rings on the alternator's armature was set at a fixed value, and as the alternator was driven to various speeds, constant volts-per-hertz was generated. This arrangement could be used to drive multiple motors for synchronized conveyor lines, at variable speeds, with only one DC motor to maintain.

With the advent of the Silicone Controlled Rectifier (SCR) in the 1960s, solid state controls were introduced to control DC motors, replacing motor-generator sets. In the following decade the SCR drive was refined to become a very reliable control system for the DC motor. In the meantime, several companies began working on designs that would use the SCR to control AC motors with variable frequency.

To provide variable frequency, the constant line voltage and frequency had to be *converted* to DC, and then *inverted* to produce a variable voltage and frequency. This required a much more complicated control system than the single conversion of the DC drive. Since the AC motor had no way of directly controlling armature current, it had to be approached with an indirect control concept.

> A DC drive is like having a mule that you can whip and force him to pull the load or break the traces. On the other hand, an AC drive is like hanging a carrot 2 feet in front of his nose and he will follow it as long as he can smell it. If it is 2 feet, six inches away, he can't smell it and stops—if it get too close, he eats the carrot, and everything stops! In other words we can force a DC motor directly, but we have to make conditions right for the AC motor so it will follow indirect controls.

In the late 1970s several companies introduced variable speed AC drives based upon SCR technology. One of the first successful drives was called the Variable Voltage Input (VVI) inverter, sometimes called a six-step. (Note that "inverter" has become a generic name for VFDs, even though only the output stage is really an inverter.) Larger drives were called Current Source Inverters (CSI)

Earlier in Figure 4-3, we showed how a rotating field is induced in the motor by applying 3 phase (3-wire) power to our 3-circuit winding. We might say that the 3-phase source is our commutator. Since we have a 2-wire, DC bus driving a 3-circuit winding we have to set-up a switching sequence to produce DC pulses that simulate the 3-phase rotating field. The switching devices become our output commutator. Figures 5-6 and 5-7 show the principles of this arrangement.

Since there is a certain amount of reluctance in the magnetic structures, the DC pulses will tend to be smoothed out somewhat and the motor will see an acceptable wave form similar to Figure 5-8. The irregularity of the wave form will increase the motor heat by 10 to 15%. All motors will run somewhat hotter on inverters than across the line, so a motor with a service factor of 1.15 should always be used to provide more thermal capacity, or a motor rated for inverter service.

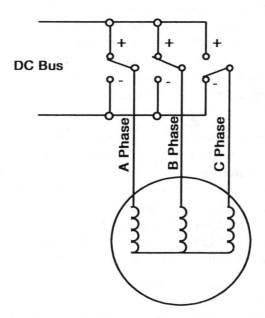

Figure 5-6 Connecting 3-Phase Winding to DC Source

Motor Response to Variable Frequency

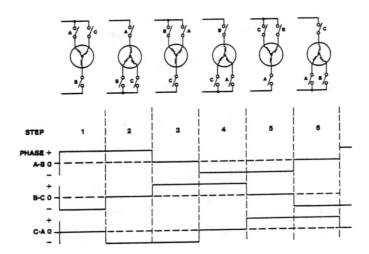

Figure 5-7 Simulating 3-Phase from DC Source

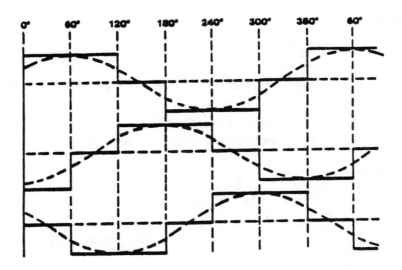

Figure 5-8 3-Phase Motor Response to Square Wave Input

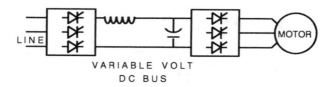

Figure 5-9 Variable Voltage Inverter (VVI)

VARIABLE VOLTAGE INPUT (VVI)

The VVI drive uses a three phase SCR bridge, like a DC drive, for the "front-end". This energizes the *variable voltage* DC bus to provide the voltage level required to maintain constant volts-per-hertz, matching the output frequency. The output stage also uses an SCR bridge or GTOs (Gate Turn-off). A GTO is a type of SCR that has a special gate that can turn the device off without waiting for a zero-crossing, using a burst of reverse current. Figure 5-9 shows the basic scheme of the VVI drive.

Advantages:

- This drive can be tested without a motor to read changing output frequencies and voltages.
- It has a fairly wide frequency capability from zero to 240 Hz range.
- It can drive multiple motors, within the current limit of the drive, and in general has a fairly simple construction.

Limitations:

- Source voltage fluctuation of more than 5 to 10% will cause it to fault.
- The power factor varies similar to a DC drive, and can be very low at low speeds.
- At low speed the drive will "cog", creating stresses on shafts and keyways.
- It cannot "ride-through" brief power loss, because with a variable voltage bus, one cannot use it to sustain controls.
- The drive will reflect current harmonics back to the line that can disturb other electronic equipment.

Motor Response to Variable Frequency

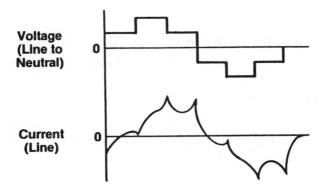

Figure 5-10 VVI Wave Form

Figure 5-10 illustrates the wave forms this system produces. An isolation transformer should be used to help trap the current harmonics.

CURRENT SOURCE INVERTER (CSI)

The CSI drive is constructed much like the VVI, in that it uses an SCR front end, and has a variable voltage DC bus. See Figure 5-11. The main difference in construction is in the large inductor in the DC bus. The mass of this inductor may equal the size of the motor. The motor must be matched to the drive, so multiple motors are not possible. The drive is really a big current regulator and one can short-circuit the secondary without a fault. In fact it cannot be tested for voltage and frequency output without a connected load.

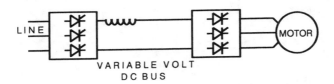

Figure 5-11 Current Source Inverter (CSI)

Advantages:

- It has relatively high efficiency.
- It is the only solid-state drive that is line-regenerative without adding equipment.
- Good short-circuit protection
- Capable of synchronous transfer (Picking up a running motor and operating it at variable frequency).

Limitations:

- Source voltage fluctuations of 5 to 10% may cause faults.
- Speed range is limited to maximum of about 66 Hz.
- Power factor varies with speed, like VVI.
- Cannot operate multiple motors.
- No ride-through, like VVI—variable voltage bus.
- Requires motor matching.
- Creates voltage harmonics.
- Cogs at low speed, so working range is limited to about 15 to 66 Hz.

Figure 5-12 shows the typical wave forms produced by this drive. Note the current interruptions that cause cogging. There is cogging over

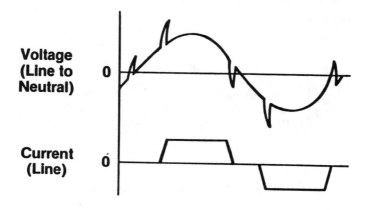

Figure 5-12   CSI Wave Form

the entire operating range, but system inertia usually absorbs it above about 15 Hz.

This system should also have an isolation transformer. While the transformer may attenuate the harmonics to certain extent, voltage transients can still pass through it.

PULSE-WIDTH MODULATION (PWM):

In the early 1980s, technology advances developed the power transistor that could support small motor loads. The transistors could be turned on and off very fast, leading to the PWM technology that could produce a current sine wave that imitated line current very closely. This reduced the motor heating, and increased system efficiency. Larger PWM drives, 100 to 1500 Hp (1990), originally used GTOs in the output stage, but as transistor capacity is growing, they are displacing the GTOs.

The PWM algorithm was originally controlled by special LSI (large scale integrated) chips and most adjustments used analog potentiometers, jumpers and dip switches. Present day drives use microprocessors and digital, programmable parameters.

The PWM drive differs from VVI and CSI construction in that it uses a diode front end and a *constant voltage* DC bus. See Figure 5-13. Bi-polar transistors for each output phase are connected to the DC bus, with center-tap going to the motor terminal. Alternately, each transistor is pulsed with a varying pulse-widths for one-half of each cycle. Each positive and negative pulse group represents one cycle. By changing the length of each cycle, different frequencies can be attained, maintaining the constant volts-per-hertz ratio, with an accuracy of 0.01 Hz.

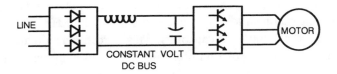

Figure 5-13 Pulse Width Modulation (PWM)

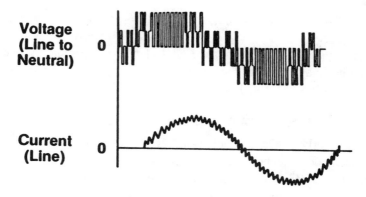

Figure 5-14 PWM Wave Form

Figure 5-14 is a representation of wave forms one might expect. Obviously, the pulses will be much denser than shown in most cases. Pulse density is greatest at the center of the pulse group so that the average power simulates the power pulse from a true sine wave. This wave form produces less motor heat than the VVI and CSI systems.

One of the common complaints is the objectionable noise created in the motor by the pulse rate—sometimes called "carrier frequency". To combat this, most present day drives use a special transistor called an IGBT (Isolated Gate, Bi-polar Transistor) that can be operated at higher frequencies, above the audible limits for the human ear. Standard "Darlington" transitors operate at about 2.5 kHz which is the audible range. (Note: If an output line reactor is installed in a 2.5 kHz system the noise will usually move from the motor to the reactor.)

The IGBTs can be adjusted to different carrier frequencies up to about 20 kHz. When operating at 16 kHz, the VFD will have about 20% more internal heat loss than one operating at 2.5 kHz, often requiring derating to get full load capability. This high frequency also causes line disturbances and insulation problems that require special consideration that will be discussed later under the topic of "harmonics".

PWM Advantages:

- Diode front-end gives constant power factor. (Looks like a battery charger to the line.)
- Wide frequency range to 240 Hz or higher.

- Highest efficiencies (up to 98%).
- Can ride-through power loss from 2 Hz to as much as 20 seconds.
- Can withstand wider line voltage fluctuations.
- Capable of common bus regeneration if multiple inverters are connected to a common DC bus. Often used for transit-bus trolley service.
- Isolation transformers are optional.
- Constant power factor at all loads and speeds.
- Full operating torque available near zero speed.
- No cogging.
- Multi-motor operations
- Many control options, PLC, etc.

Limitations:

- Extra equipment required for full line-regeneration.
- Special equipment may be needed to correct harmonics problems when using very high frequency IGBT drives, such as line and load reactors.
- IGBT drives require motor insulation rated to withstand 1600-2000 volt spikes.

As the microprocessor technology has advanced, many drives now have built-in logic to perform PID functions, pump staging, timing, etc. Some have interfaces that make them appear as an addressable I/O (input/output) rack to a system PLC (Programmable Logic Controller).

SCALAR VS VECTOR CONTROL

Scalar

The most common PWM AC inverter control, above, provides an output frequency that is scaled to the output voltage. In other words it is a "voltz-per-hertz" control scheme. Many scalar controls have a number of programmable functions, giving them the sophistication that can provide customized V/Hz curves to match required performance. The control is digital, so it is not subject to the drift associated with analog systems.

The control algorithms may be provided by an LSI (Large Scale Integration) chip or micro-processor(s). The preciseness of the control depends upon the microprocessor. An eight-bit microprocessor will provide frequency regulation within 0.5 Hz. A sixteen-bit microprocessor can control within 0.01 Hz.

True torque control, and line regeneration are difficult to attain with standard scalar drive controls. They are basically open-ended speed-control systems with no external feed-back. This design cannot compete with the capabilities of a demanding DC control system, because it cannot effectively control torque.

Vector

With the introduction of Vector control, the AC motor can now compete with nearly any DC application. Vector control may also be called Field-Oriented Control, Flux Control (Indirect), Torque Control, and Slip Control. Several manufacturers offer proprietary vector control systems. The following is a brief description to show the contrast between scalar and vector drives. The complexity of vector drives goes beyond the "fundamentals" reviewed in this book, but we will give you some insight into the basic concept.

Many vector controls require a feed-back such as tachometer, resolver, or encoder to provide shaft speed and position feed-back. (Some manufacturers have developed vector drives that can now operate without these mechanical feed-back devices.) Using the feed-back with a mathematical motor model, and current vectors, it is possible to determine and control the actual speed, torque, and power produced by the motor on a continuous basis.

Some drives use a custom-built motor which has known parameters. In this case a "personality module" may be plugged into the control to provide the mathematical model. The most sophisticated drives have "auto-learn" capabilities which allow them to self-calibrate themselves to match the motor. This is especially useful in the event a motor is rewound or replaced. The drive logic will require the following parameters to create a mathematical model of the motor.

— Stator resistance (cold and hot)
— Rotor resistance (cold and hot)
— Slip
— Power factor

Motor Response to Variable Frequency

- Stator voltage
- Stator current
- Stator frequency
- Saturation of the iron
- Stator inductance
- Rotor inductance
- Magnetizing inductance

Figure 5-15 shows one method depicting vector control. Looking at a single phase sine wave, you will see that at 30°, the real voltage is 50% of maximum. At 60°, it is 87%, and reaches 100% at 90°. This is shown as a time representation.

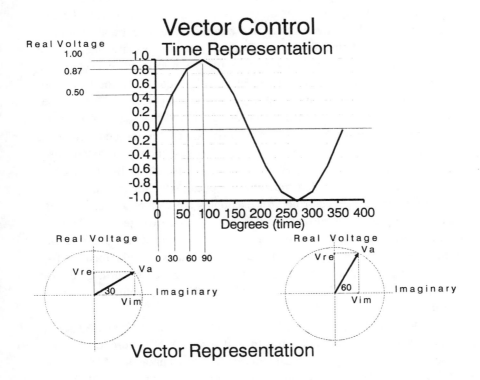

Figure 5-15 Vector Definition

For a vector representation, we visualize an unchanging voltage magnitude that is being rotated. At any point in the rotation, the voltage is the vector sum of real voltage and imaginary (excitation) voltage. Real voltage is shown as a vertical component, and the imaginary voltage is shown horizontal. The real voltage is always the sine of the angle of displacement. e.g. SIN 30° = .50; SIN 60° = .87.

While we are showing only one phase, remember that there are three phases involved, displaced by 120°.

Current vectors are made up of magnetizing (horizontal) and torque producing (vertical) vectors. To continually control the current, we must compensate for the angle of voltage displacement (power factor), and the time constant (magnetic and thermal delay) of the rotor. These are dynamic conditions that vary with different load demands, and temperatures.

The system first performs a transform function which places the current vector on the voltage vector and establishes new real and imaginary vectors, compensating for power factor. These vectors are then rotated forward to anticipate and compensate for the rotor time constant. The required torque vectors are then created with reference to the calculated real (power) and imaginary (magnetizing) vectors. (Figure 5-16.) In other words, if we know the vector angle required, and have control of when to fire the transistors, we can continuously control the torque (current) in the system. This is a continuous calculation responding instantaneously to all changing parameters. By firing rapidly ahead and behind a given position, we can have full torque at zero speed, for holding loads.

In essence, the vector control decouples the excitation current from the torque producing current, similar to a DC drive where the field excitation current is separated from the armature current. Once this relationship has been established, further control becomes straight forward like DC drives. This is an over-simplification of the process, but it gives one the general idea of the complex computation that is going on continuously in a vector control system.

One manufacturer's specification shows that a 40 rpm speed step will rise in 100 msec and settle in 220 msec. A 50% torque step will rise 25 msec and settle in 200 msec. Static accuracy is .01% speed, .02% torque. The motors usually have separate motorized cooling fans so they can operate at full torque at any speed.

Motor Response to Variable Frequency

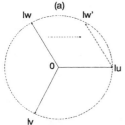

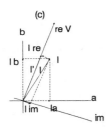
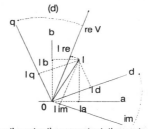

(a) 3-Phase current vectors
Add vector Iw to Iu for single phase calc.
Sum is I in (b)

(b) I = Iu + Iw'
Transform I to Ia & Ib
I represents stator current
Ia-Iw' = Torque vector for current I

(c) Voltage vector positioned relative to stator current vector (Power factor)
Real axis I placed on voltage vector I'
Imaginary axis im is perpendicular to it.
Transform vector I into real & Imag. axes.

(d) Using the rotor time constant, the vector I is transformed into the rotating axes of the rotor.
Iq is new torque producing vector.
Id is the new magnetizing vector.
Iq, Id will produce I with this Pf and rotor time constant.

Figure 5-16 Vector Representation

Advantages:

1. Precise control of motor speed, torque and power, even in constant voltage range.
2. Quick response to changes in load.
3. Quick response to changes in speed command.
4. Quick response to changes in torque command.
5. Can provide 100% torque at 0 speed. (Can hold a load).
6. Torque sharing among motors.
7. Smooth low speed operation.
8. Smooth forward/reverse and start/stop operation.
9. Less mechanical stress due to jerk.

Typical applications:

1. Helper drives, torque sharing, tension control
2. Dynamometers, crane, hoists
3. Traction/propulsion drives
4. Winders, unwinders
5. Paper machines
6. Hard to start loads, such as extruders, positive displacement pumps.
7. Loads where load torque and moment of inertia vary wildly.
8. Need to start a spinning load in either direction.
9. Any application where maximum probability of success is required.

The vector drives are more complex, and expensive than the scalar systems. In general, a good vector drive can match DC performance with a lot less motor maintenance, and maintain high power factor at all speeds and loads. Regenerative front-ends are also available for those applications where high regenerative inertias must be controlled.

Some manufacturers utilize common power structures for scalar and vector drives, but provide optional "mother-boards" to suit the required operation. This commonality helps reduce the spare parts inventory.

With the vector control capabilities, many previously specified DC applications are now being replaced by the new AC technology.

Chapter 6

Harmonics

With the proliferation of many new electronic power products such as AC and DC motor drives, computers, switching power supplies, etc., the study of harmonics has become a serious topic. A power distribution system may start having heat problems in transformers, motors, wiring terminals and distribution conduits that seem to be unexplainable. There may be capacitor and motor winding failures. The cause of the problem goes undetected because your typical RMS clamp-on ammeter does not show the excess harmonic current.

In the past most electrical equipment operated directly on the *fundamental frequency* of 60 Hz, so harmonic content was negligible. These are considered *linear* loads. Electronic devices utilize many silicone controlled rectifiers (SCRs) and transistors that draw power from the line in short interrupted bursts at frequencies that do not match the 60 Hz source. These loads are considered to be *non-linear*.

The bursts of current from non-linear loads super-impose *harmonic* frequencies on the fundamental frequency, resulting in the distortion of the current and voltage sine waves. These harmonics are reflected "upstream" and will have an impact on the distribution system. Their total effect is related to how much current capacity that the power source can supply, compared with the total harmonic currents produced. Harmonics that have little effect on a "stiff" power system could totally disrupt a small emergency power plant.

Harmonics are currents or voltages with frequencies that are integer multiples of the fundamental power frequency. For example, if the fundamental power frequency is 60 Hz, then the 2nd harmonic is 120 Hz, the 3rd is 180 Hz, etc.

CLASSIFICATION OF HARMONICS									
Name	F	2nd	3rd	4th	5th	6th	7th	8th	9th
Frequency	60	120	180	240	300	360	420	480	540
Sequence	+	-	0	+	-	0	+	-	0

Harmonics that have a positive (+) sequence have a forward rotation as applied to motors and will increase heating due to skin effect and eddy currents.

A negative (-) sequence produces a reverse rotation (torque) that causes excess heating in motors as it partially cancels the effects of the fundamental frequency. The 5th and 11th harmonics are usually the worst offenders for negative sequence problems, and are generated by SCR switching, as used in DC, CSI and VVI drives.

If the sequence is zero (0), there is no effect on rotation, but the neutral of a 3 phase, 4-wire system will carry 100% of those harmonic currents and create heat problems. The neutral current often exceeds that of the phase conductors. The 3rd harmonic current has been known to burn out neutral conductors, so many codes now require an over-sized neutral conductor in buildings that have a large computer load. If the service transformer has a wye-delta configuration, the 3rd harmonic current will be trapped in the delta winding, and become a circulating current that will add to the fundamental current and become a source of additional heat. Switching power supplies, used in computer equipment, are a source of high 3rd harmonic currents. See Figure 6-1.

Even numbered harmonics cancel out, and have no appreciable effect on the system.

The typical RMS meter will read fundamental current quite accurately, but cannot sense the harmonic currents. In fact if one is reading a square-wave load, the meter may read 10% high, but when looking at a distorted wave such as produced by a variable frequency drive (VFD) it may be 40% low. For accurate ammeter readings, one should consider getting a true RMS meter such as Fluke 87DDM, or equal, that has a frequency band-width of 2 kHz or more.

Harmonics

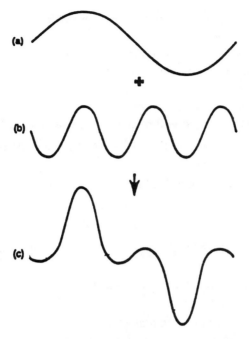

Electrical Waveshape Graphics showing how sine waves of different harmonic frequencies combine to form a distorted waveshape at the fundamental frequency. (a) A sine wave of current at the fundamental frequency of 60 Hz. (b) A sine wave of current at the 3rd harmonic of 180 Hz. (c) The combination of (a) and (b) resulting in a distorted waveshape at 60 Hz.

Figure 6-1 Third Harmonic Distortion

SUMMARY OF HARMONIC PROBLEMS IN COMMERCIAL BUILDINGS

- Phase conductors, heating from skin effect
- Conduit, heating from skin effect, eddy current, and conductor heat
- Circuit breakers
  — False tripping due to heat and high harmonic frequencies
  — Peak-sensing electronic circuit breakers respond to peak wave form and trip at prematurely low current levels because harmonic current is higher than normal wave form.

- Neutral conductors, high current from 3rd and 9th harmonics (zero sequence).
  — Harmonics add in neutral rather than cancel.
  — Often 130% of RMS phase current.
  — No circuit breaker in neutral to protect circuit, often too small conductor.
  — <u>Neutral current</u> will read about the same on all meters.
- Excess voltage between ground and neutral at receptacles, due to harmonic load
- Neutral bus bar overload (heat) from zero sequence harmonics
- Neutral lug over-heating from overload
- Panel steel, eddy current heat, vibration, noise
- Transformer heating from eddy currents and circulating harmonic current
- Induction motors, heat from eddy currents and negative sequence currents
- Inaccurate readings on ammeters, kW and kVAR meters
- Telecommunications, adds noise to phone, especially in audible range >500 Hz
- Standby generators may not have capacity to absorb total harmonic load
  — Overheat from excess current
  — Regulation may suffer because of false zero-crossings from harmonics
- Power factor correction capacitors may fail at resonant frequencies

When harmonics become a problem, it is wise to consider adding filters to trap the worst offending frequencies. This requires some expertise, and should be assigned to people who have experience in harmonics containment. Most specifications require that total voltage distortion should not exceed 5%, and current distortion should be less than 20%. Power utilities are sensitive to this because of the excess loading of their substation equipment. Transformers frequently have to be de-rated to prevent over-heating from the high core losses that the harmonic currents produce. Many distribution transformers now have a "K-rating" which is a de-rating factor applied where loads have high harmonic content.

Harmonics

There are a number of engineering firms that specialize in correcting harmonics problems. They typically make an analysis of the plant power system, and specify the filtering equipment necessary to balance out each troubling harmonic. The study and control of harmonics is a subject that is very specialized, and beyond the intended scope of this book. As drives become a greater share of the plant load, one must be aware that harmonics will become an increasing source of power and distribution problems.

AC drive manufacturers that supply IGBT power supplies offer optional line inductors and accessories to help limit the harmonics that are created. The installation of the optional protective equipment will pay-off in increased insulation life for the affected motors. To minimize harmonics and increase efficiencies, operate IGBT drives at the lowest carrier frequency for acceptable noise suppression. Above 4 kHz, derating is often required. It has been predicted that by the year 2000, 60% of our electrical loads will be electronic, so harmonics will continue to be a concern when considering future power quality.

Chapter 7

PWM Drive Parameters

The following describes, in general terms, the operating sequence of a typical PWM drive. Since most modern drives now have similar programmable parameters, we will assume that the more common parameters are present. A *parameter* is a programmable function that has adjustable limits, e.g., acceleration time, overload limits, etc. It can have a read and write function, or a read only output.

Before a motor can be started, there are usually five conditions that must be met:

1. A momentary "true" input for the "start" function. This may be a normally open push-button, or logic from an interface with programmable controllers or computer devices.

2. A momentary "false" input for the "stop" function. This may be a normally closed push-button, or logic from interface with other controls. It is desirable to have at least one "hard-wired" input to allow over-ride in the event of other logic failure.

3. A momentary "false" input for the "enable" function. This is usually a conditional input proving that all conditions have been met to allow a safe start. It may be internal to the drive, and/or from an external source.

4. A speed reference to set the operating frequency. This may be a simple potentiometer, an analog signal, a pulse train input or a bit pattern, depending upon the drive design and interface equipment.

5. An auxiliary maintained "true" input or maintained closed contact to permit the drive to start, run, or jog. This might be an external motor thermostat, overload, or machine safety contact that can disable the drive. If this function is tripped, it will usually generate an identifying fault code in the drive.

STARTING THE DRIVE

With power on, DC bus energized, and the above conditions met, one can initiate a "start" command. The main control section will check for system faults, and if clear, instruct the base driver board to energize the power transistors for the frequency and volts-per-hertz commanded by the speed reference. The normal time it takes to reach commanded motor speed will be controlled by the rate programmed in the acceleration parameter. Current is continuously monitored by hall-effect sensors, transistor saturation, or other means. If current exceeds the over-current parameter settings during this period, it will over-ride the acceleration settings, and either extend the acceleration time, or trip the drive off. Extending the acceleration time will usually prevent nuisance trips when starting.

As the frequency and volts rise above the rated slip-frequency of the motor, current (torque) will increase against the connected load, until the motor breaks away and accelerates to the commanded speed. The typical constant torque drive can produce 150% break-away torque depending upon how close the inverter matches the motor.

Special cases

Some drives have parameters that can be set to compensate for the different harmonic reflections from special motors. If unexplained over-current faults occur, it may be necessary to make a parameter change to accommodate permanent magnet, or synchronous induction motors. It may also require a special volts-per-hertz adjustment for motors of foreign origin or those with unusual frame sizes. Very high speed router motors usually have special volts-per-hertz requirements, and may also require output inductors to compensate for the low iron mass in the motor. In the event that the parameter adjustments can not compensate for the odd motor, a larger inverter may be required.

STOPPING THE DRIVE

Most drives can be programmed to coast-to-stop, ramp-to-stop, or possibly brake-to-stop when the stop command is initiated.

When set-up for *coast-to-stop,* the motor is de-energized and its stopping time will be determined by the inertia and friction in the system. It is an uncontrolled stop.

Ramp-to-stop directs the base driver to reduce frequency to zero with a controlled deceleration rate to power the motor to a stop. In this case, the energy is fed back through the inverter section to the DC bus. The drive must be able to absorb the energy required to stop the motor. If the internal losses in the drive are not high enough to prevent an over-voltage fault, the deceleration time may have to be extended, a deceleration frequency hold programmed to clamp the excess voltage, and/or a dynamic brake kit may be added. The dynamic brake control senses the DC bus voltage, and when a predetermined threshold is reached, it activates a chopper control to bleed excess voltage to a resistive grid bank. The excess energy is dissipated as heat in the resistors.

Brake-to-stop injects controlled DC current into the motor winding during a preset time period to bring the motor to rest. The rotating field is now stopped and locked, while the rotor is still spinning. The motor is brought to a stop by the counter-torque this condition generates. The squirrel-cage rotor dissipates the energy within itself as heat. Repetitive braking can damage the motor if the heat buildup becomes too great. As a rùle, this mode of stopping is useful for braking a lightly loaded motor, but is not meant to control high machine inertia.

REVERSING THE DRIVE

Drives typically have a parameter that enables/disables the reverse function.

Reversals are usually controlled by a contact that is open for Forward, and closed for Reverse. When reverse is commanded, the drive reduces speed to zero and then comes up to speed in the opposite direction using the Accel/Decel settings of the drive. The base driver board merely changes the switching rotation of the transistors to accomplish the reversal.

Step-changes in frequency, or commands for pre-set speeds also rely on Accel/Decel settings, with current protection, to make smooth speed transitions. If over-voltage faults occur during changes from a high speed to a lower speed, it is an indication that the deceleration rate is too fast, and/or a dynamic brake needs to be installed.

RUNNING CONDITIONS

The inverter often has a protective feature called Momentary Overload Protection Circuit (MOPC). This feature monitors the current at all times, and if the load increases to 150%, (or other preset value), the drive will respond by reducing frequency and voltage to limit output current to this value. Once output current is reduced below this value, the drive will ramp back to command frequency. This gives the drive the capability of riding through reasonable overload conditions without nuisance fault trips.

When an overload condition exists, the drive modulates output current by going into and out of MOPC. Logic senses this as a non-continuous MOPC condition. If output current continuously exceeds the value set, for perhaps four seconds, the drive will fault, initiate a stop sequence, and display the appropriate fault code.

A separate overload current parameter may be available which simulates the typical thermal overload heaters required by the NEC & UL. When current exceeds the setting, timing begins which is inversely proportional to the level of current. If the overload times out, the drive will fault, coast to stop, and display a fault code. The overload range may be limited by MOPC settings. It should be noted that a thermal overload device will not protect a motor operating at slow speeds because the motor can burn-out at rated current when the cooling fan is ineffective. The best protection is to install a thermostatic device in the windings of the motor which either provide an alarm, or trip the system off when the windings reach a critical temperature limit.

Slip compensation is a parameter which is available in the more sophisticated drives. It is used to minimize the change in motor speed due to changing loads. As the current increases, this parameter can be adjusted to slightly increase the commanded frequency, compensating for the additional slip the load is inducing.

A fan application will often display mechanical resonance at certain critical speeds. This is not only noisy, but can be destructive if allowed to operate continuously at that speed. Skip frequency, and skip frequency band-width parameters, are usually available to prevent the drive from settling on critical speeds. The skip frequency is set for the center of the resonance zone, and the band-width is set to identify the total range of the skip frequency. In other words, the critical speed might be at 43 Hz, but may require a total band-width of 6 Hz to reach a quiet zone. When accelerating, the drive would reach 40 Hz, and would not accelerate further until the command reaches 47 Hz. Decelerating, the drive would settle at 47 Hz and skip to 40 Hz as the command decreased. Usually, up to three skips can be programmed, with the required band-width settings.

It should be remembered that current is the critical value that influences an AC drive's behavior. The drive primarily uses current to reflect the motor torque, and slip. Bus voltage is used to sense regenerative conditions. The microprocessor or LSI chip is continually updated and makes corrections to compensate for dynamic changes. Parameters are used to establish acceptable limits the motor and drive are expected to tolerate.

Precautions

Drive logic should always be used to start, stop, and control the drive. If an existing contactor is left in the circuit when a drive is installed, it should only be opened or closed when the drive is in a STOP condition.

A contactor installed ahead of the drive can be used for an isolating disconnect switch. A contactor installed between the drive and motor should never be operated when the motor is running. When opened under load, there is possibility of an arcing spike that can destroy a transistor. Closing the contactor against an operating drive will submit the drive to a cross-the-line current surge (600-800%) that will trip the drive off-line, and stress the transistors. The only exception would be when multi-motors are operating from a single inverter that has enough current capacity to carry running motors, and also the in-rush current of the single motor.

PARAMETERS

Parameters serve two functions in drive systems. They provide logic access to set (write) constants for performance control, and also return information (read) for monitoring the performance. Most drives manufactured in 1992, or later, have programmable parameters. Parameters may number from about 30 to as high as 2000, or more, depending upon the sophistication of the product.

Figure 7-1 Photo, AC Drives, Courtesy of Allen Bradley Co. Inc, Milwaukee, Wisconsin

The more common parameters are defined in the following pages. They may be defined differently by various manufacturers, but the basic functions will be the same. *Read-only* parameters are used to show status, and are useful for trouble-shooting and monitoring. Many Allen-Bradley, Bulletin 1336 drive parameters, available in 1990 and later, are listed as examples. An asterisk (*) marks the parameters that will be common to most drives.

Parameter 0, Parameter Mode*

Read Only

Displays parameter programming status. A decimal point in the display indicates that programming is enabled. If decimal point is not displayed, the parameters are in a "read only" state.

Displays running frequency.

Displays fault code if system faults.

PWM Drive Parameters

Parameter 1*, Output Volts
Read Only
Displays running output voltage to the motor.

Parameter 2, Output Current
Read Only
Displays percent of rated drive output current drawn by the load, calculated from internal information. Manufacturers' conversion tables are available, for conversion of percent current to amperes, for each of their drives.

Parameter 3, Output Power
Read Only
Displays percent of rated drive power drawn by the load, calculated from internal information. Manufacturers provide conversion tables that are available to convert to HP or kW.

Parameter 4*, Last Fault
Read Only
Displays the Fault Code for the last drive fault that has occurred, even after a reset. It is updated each time a new fault occurs. Some drives maintain a historical record of the last series of faults.

Parameter 5, Frequency Select #1
Read and Write
Selects source of frequency reference
    0 - Local panel potentiometer
    1 - 0-10vdc source
    2 - 4-20ma source
    3 - Pulse train source
    4 - Serial input source
    5 - Remote speed potentiometer
Displays frequency source presently in control.
    0 - Local panel potentiometer
    1 - 0-10vdc input
    2 - 4-20ma input
    3 - Pulse train input
    4 - Serial input
    5 - Remote speed potentiometer
    6 - Jog

7 - Preset frequency #1
8 - Preset frequency #2
9 - Preset frequency #3

Parameter 6, Frequency Select #2
Read and Write
Identical to parameter #5, allowing alternate source to be used.

Parameter 7*, Accel Time #1
Read and Write
Sets the time in seconds required to reach the maximum programmed frequency (Par19), with maximum frequency commanded. This time may be modified by selection of alternate Accel rate #2 (Par 30), MOPC (Par36), and Dwell (Par43) settings.

Parameter 8*, Decel Time #1
Read and Write
Sets the time in seconds required to ramp down from maximum frequency (Par19) to zero hertz. This time may be altered by alternate Decel rate #2 (Par31), or if Decel Frequency Hold (Par11) is active with a high DC bus voltage.

Parameter 9*, DC Boost **
Read and Write
Sets DC boost voltage at low frequency to accommodate special starting torque conditions. This value should be kept as low as practicable to prevent excess heat in the motor. Using a setting of 11, and Parameters 48, 49, 50, a custom volts-per-hertz curve can be attained. See Figure 7-2.
**This is sometimes called "IR Comp."

Parameter 10, Stop Select
Read and Write
Select Coast-to-Stop (0), DC Brake-to-Stop (1), Ramp-to-Stop (2).
0 - On stop command, frequency shuts off and motor coasts to stop. DC Hold Time (Par12) and DC Hold Volts (13) must be set to zero.
1 - On stop command, frequency shuts off and DC Hold voltage (Par13) is applied for the time set in DC Hold Time (Par12), to reduce stop time. Heat is dissipated in the rotor, depending upon the frequency of the stops and the DC voltage applied. This should not be used to control high inertia loads. Excess voltages may also be generated during this cycle.

2 - On stop command, frequency ramps down at a rate dependent upon the selected Decel parameter. If values have been programmed in DC Hold Volts (Par13) and DC Hold Time (Par12), braking cycle will be applied at zero hertz. This reduces the tendency for the motor to coast at zero hertz.

Parameter 11, Decel Frequency Hold
　　Read and Write
　　When activated by selecting (1), the parameter will clamp the decel rate during the time DC bus voltage exceeds 110% of nominal. This extends the decel time to prevent Over-Voltage trips from regenerative deceleration rates that exceed limits. This parameter may need to be turned off (0) if line volt surges raise DC bus voltage above 110% of nominal, to prevent it from over-riding a command to ramp down.

Parameter 12, DC Hold Time
　　Read and Write
　　Sets the number of seconds that DC Hold Voltage is applied. Stop Select (Par10) determines how this timing is applied. Time should be set to Zero if braking is not used.

Parameter 13, DC Hold Volts
　　Read and Write
　　Sets the DC Hold Voltage that is applied during the DC Hold Time (Par 12). Set at lowest functional value to prevent motor overheating. The heat from DC braking is dissipated in the squirrel cage rotor, and can damage stator windings also, if temperatures become excessive.

Parameter 14*, Auto Restart
　　Read and Write
　　Determines restarting sequence upon line loss fault. If set to (0), restart will require cycling the stop, or power input. If set to (1), Bus Undervoltage fault is inhibited. If Power Fault Par (40) is set to (1), an automatic restart will happen—if a maintained start input is present, such as a pressure switch, etc.
　　Auto restart may be a dangerous condition if an unannounced restart occurs where people might be injured.

Parameter 15, Unused at present
　　Read and Write

Parameter 16*, Minimum Frequency
    Read and Write
    Sets the minimum frequency required with potentiometer set at (0), voltage input at (0), or current input of (4ma).
    This can be adjusted as an offset to allow operation between two limits, without going to zero except to stop. Jog frequency (par24) and Dwell frequency (Par43) can over-ride this value.

Parameter 17*, Base Frequency
    Read and Write
    Usually set at motor nameplate value or (60). It establishes the maximum frequency for the constant volts-per-hertz range of operation. The volts-per-hertz slope can be modified by settings of DC Boost (Par9), Base Volts (Par18), Start Boost (Par48), Break Frequency (Par49), and Break Volts (Par50). See Figure 7-2.

Parameter 18*, Base Volts
    Read and Write
    Usually set at motor nameplate value or (460). It establishes the maximum voltage for the constant volts-per-hertz range of operation. Volts-per-hertz slope can be modified as discussed in Par (17), above. This is not necessarily the maximum voltage available, see Par (20). See Figure 7-2.

Parameter 19*, Maximum Frequency
    Read and Write
    Sets the maximum frequency that will be available when maximum speed command is given from any source. If set for less than Base Frequency (Par17), this parameter will over-ride the base frequency. See Figure 7-2.

Parameter 20*, Maximum Volts
    Read and Write
    Sets maximum available volts from the drive. It is used primarily for custom applications where the voltage at Base Frequency is less than Maximum voltage. Example: An application driving 230 volt motors to 120 Hz and 460 volts. See Figure 7-2.

Parameter 21, Local Run
    Read and Write

PWM Drive Parameters

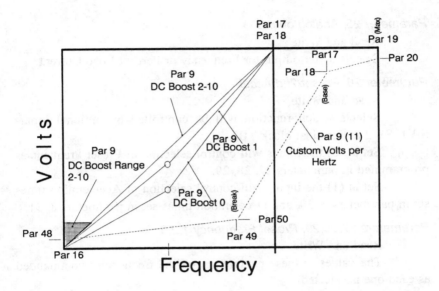

Figure 7-2 Parameters Affecting Volts-per-Hertz

Enables the local start command to work with other remote sources. Can be disabled to prevent local starts when machine is out-of-sight and dangerous starting conditions might occur.

Parameter 22, Local Reverse
   Read and Write
   Exclusively Enable or Disable local reversing push-button. Other reversing sources are Disabled/Enabled respectively.

Parameter 23, Local Jog
   Read and Write
   Enable or Disable local Jog push-button. Has no effect on other jog inputs.

Parameter 24*, Jog Frequency
   Read and Write
   Sets frequency drive will ramp to when Jog is commanded. If no other frequency is commanded, it will follow the stop mode set in Par #10. This command can be set lower than Minimum Frequency (Par16), but will be limited by Maximum Frequency (Par19)

Parameter 25, Analog output
   Read and Write
   Selects meter output for Frequency or Percent Load Current.

Parameter 26, Preset/2nd Accel
   Read and Write
   Selects which function will be controlled by optional inputs SW1, SW2 at Terminal block TB3.
   Set at (0) the inputs will control selection of Preset Frequencies programmed in parameters 27, 28, 29.
   Set at (1) the inputs will control selection of Acceleration times set in parameters 7, 30, and Deceleration times set in parameters 8, 31.

Parameters 27,28,29, Preset Frequency 1,2,3
   Read and Write
   The values in these parameters are the frequencies commanded as each one is selected.

Parameter 30, Accel Time 2
   Read and Write
   The value, in seconds, determines the time to ramp from zero to Maximum Frequency (Par19), when selected.

Parameter 31, Decel Time 2
   Read and Write
   The value, in seconds, determines the time to ramp from Maximum Frequency (Par19) to zero, when selected.

Parameters 32,33,34, Skip Frequencies 1,2,3
   Read and Write
   Sets the center point of up to 3 frequencies where the drive will not settle. (It will operate above and below these frequencies.) This is primarily used to prevent the motor from settling on a resonant frequency that might be creating vibration and noise.

Parameter 35, Skip Frequency Band
   Read and Write
   The skip frequencies (par33,34,35) plus and minus this value equals the total range of the skipped frequencies.
   If the skip frequency is set at 30, and Par35 is set for 5, the drive will not settle between 25 and 35 hertz.
   Overlapping ranges may be programmed if a large range is required.

PWM Drive Parameters

Parameter 36*, MOPC (Fold-back Current Limit)
Read and Write

The value of this parameter adjusts the Momentary Overload Protection Circuit setting of the drive in units of percent of rated output current. When output current exceeds this limit, frequency and voltage will be reduced to limit the current to this value. It will ramp back to command frequency if the current drops below this limit. If the condition persists for more than four seconds, the drive will fault, initiate a stop sequence, and display Fault Code F06.

Parameter 37, Baud Rate
Read and Write

Sets baud rate to communicate with serial devices to 2400 (0) baud or 9600 (1) baud. Power must be recycled whenever baud rate is changed.

Parameter 38, Overload Current
Read and Write

Used with single motor drives to provide overload relay function equal to that required by UL and NEC. When value is exceeded, timing starts which is inversely proportional to the excess current. This simulates an overload heater. The value is based upon a percent of drive rating, so it should be determined using instruction book tables to match motor current. MOPC (Par36) can limit the overload range.

Set at 100 when using multiple motors.

Parameter 39*, Fault Clear
Read and Write

Determines sequence of events required to clear a fault. Set at (0), fault requires removing power, letting the DC bus discharge, and reapplying power to reset. Set at (1), fault may be cleared by cycling power or toggling the stop circuit. Auto Restart (Par14) will affect the sequence to reset after a fault. Power Fault (Par40) will affect the sequence after power interruption.

Parameter 40, Power Fault
Read and Write

Selects response sequence when power interruption occurs. Set to (1), after interruption, drive will operate until bus voltage has dropped 15% below its nominal value. At this time the output shuts off to conserve DC bus power.

Logic will be retained as long as the bus voltage stays above 388 VDC. If power is restored and bus voltage raises above the -15% value, the drive will restore output power. If voltage falls below 388 VDC, Bus Undervoltage Fault (F04) will be displayed.

Set to (0), after interruption, Input Power Loss Fault (F03) may be displayed. In addition to the above sequence, when bus voltage drops below -15%, and is above 388 VDC, a 500ms timer is started. If the timer times out, F03 will be displayed and drive will trip and shutdown. If power is returned before 500ms, and voltage did not fall below 388 VDC, the drive will resume operation and reset the timer.

Parameter 41, Motor Type
Read and Write

Current characteristics for Synchronous motors are different from Induction motors. Since they operate at zero slip, the rotor current is not summed with the stator current, and the drive must be adjusted for the condition, or possibly oversized to absorb the different reflected harmonics.

Set at (0) matches the drive to a standard induction motor.

Set at (1) matches the drive to a synchronous reluctance motor.

Set at (2) matches the drive to a synchronous permanent magnet motor.

Parameters 10,42 are dependent upon Parameter 41.

DC Brake-to-stop (Par10) must NOT be set to (1) when this is set to (2) as the braking flux may demagnetize the permanent magnet rotor.

Slip Compensation (Par42) must be set to (0) when this is set for (1) or (2) as these are zero-slip motors.

Parameter 42, Slip Compensation
Read and Write

Motor current rises as load increases the motor slip. Slip compensation is a control that senses the increased current, and adds compensating frequency to hold the motor near its rated operating speed. Speed fluctuations due to load variations can thus be reduced.

Slip compensation can be set from 0.0 to 3.0 hertz. This is added proportionally to the current change and gives speed regulation in the area of 1%. An "at-speed" circuit in the drive monitors the control and as long as the command frequency is unchanged, it functions. If a change is called for, it will resume action after the drive has ramped to the new speed.

If slip-compensation is needed at maximum speed, it may be necessary to increase the Maximum Frequency (Par19) by 2 or 3 hertz to allow "head-room" for the slip-compensation frequency.

Parameter 43, Dwell Frequency
Read and Write

The value of this parameter sets a frequency that the drive will immediately produce when a Start command is received, and Dwell Time (Par44) is not set to (0). The drive will jump directly to this frequency without any acceleration ramp. It will stay at this value for the time programmed in Dwell Time (Par44), then will ramp up or down to the command frequency along the appropriate ramp. This effectively gives the motor a "kick-start" by hitting it with a higher than normal initial slip. This value is seldom set above 6 hertz.

Parameter 44, Dwell Time
Read and Write
Value is time in seconds that Dwell Frequency (Par43) is held.

Parameter 45, PWM Frequency
Read and Write

This parameter sets the minimum carrier frequency used to generate the PWM output waveform. In the 1336 drive, between 0 and 95 Hz, the carrier frequency will be 21 times the generated output frequency, but no less. The adjustment can be made from .40 to 2.00. The value .40 represents 400 hertz, and will be used between 0 and 19 Hz of output. Above 19 Hz, the carrier will change proportionally to the drive output frequency.

Increasing PWM frequency usually reduces audible noise at low speeds, but may result in instability on lightly loaded motors. Decreasing PWM frequency improves motor stability, but increase audible noise.

Some manufacturers use Isolated Gate Bi-polar Transistors (IGBT) which can operate at much higher frequencies. They have the advantage of being quieter, but the higher frequencies stress winding insulation, and may reduce the life of older motors. The higher frequencies also may increase the heat losses in the drive by 20% to 25%.

Parameter 46, Pulse Scale Factor
Read and Write

This is a parameter used to scale a pulse train input to match the frequency desired. In this case, a setting of 64 used with a pulse train input of 3840 Hz will produce a drive command frequency of 3840 / 64 = 60 Hz.

Parameter 47, Language
Read and Write

The 1336-MOD-E1 Handheld Terminal can display text in one of six languages. This parameter selects the language to be used.

- 0 - English
- 1 - French
- 2 - Spanish
- 3 - Italian
- 4 - German
- 5 - Japanese

Parameter 48, Start Boost
Read and Write

This parameter sets the DC Boost Select when (Par9) is set at (11). Parameter 48, 49, 50 are used to construct a custom volts-per-hertz curve. See Figure 7-2.

Parameter 49, Break Frequency
Read and Write

This parameter sets an intermediate frequency below Base Frequency (Par17) if DC Boost Select (Par9) is set at (11). Parameter 48, 49, 50 are used to construct a custom volts-per-hertz curve. See Figure 7-2.

Parameter 50, Break Volts
Read and Write

This parameter sets an intermediate voltage below Base Volts (Par18), if DC Boost Select (Par9) is set at (11). If this setting is lower than Start Boost (Par48), fault (F11) operator fault will result. See Figure 7-2.

Parameters 51 through 71 are used with serial interfaces such as PLCs to accomplish full control of all functions through manipulation of bytes of information. They also provide for reading and reacting to all faults and operating information. Several manufacturers have similar communication interfaces available. Drives with many parameters often have the parameter list broken down into "pages", allowing faster parameter selection, without having to go through a single long list. It is important to consult your operator's manual before doing a lot of parameter "tampering" because manufacturers may define their parameters differently.

Chapter 8

Maintenance of Pulse Width Modulated Drives

The accompanying schematic, (Figure 8-1), illustrates the principal generic components found in PWM drives. For simplicity, items such as fuses, snubber circuits, and interface arrangements have been omitted. Each manufacturer has his own unique "bells and whistles," such as parameter programming, special inter-face provisions, sensors, etc., that will require reference to the Service Manual.

COMPONENTS

The PWM drive has three principal power sections. These are the input converter (b), the DC bus (c), and the output inverter (d). The control sections will include the base driver(s) (g), and a control/interface section (h) which may be comprised of several boards with different functions. Before dwelling on trouble-shooting procedures, we will consider the function of each of the major components.

Input Power Components, Schematic Item (a)

The input power requires a disconnecting means, and protective fuses or circuit breakers to meet normal NEC rules. These components may be supplied externally, or may be a part of the drive assembly. In some cases, input inductors, or isolation transformers may be installed to control voltage and harmonic problems. Most PWM drives can operate without isolation transformers if power is reasonably stable. Using

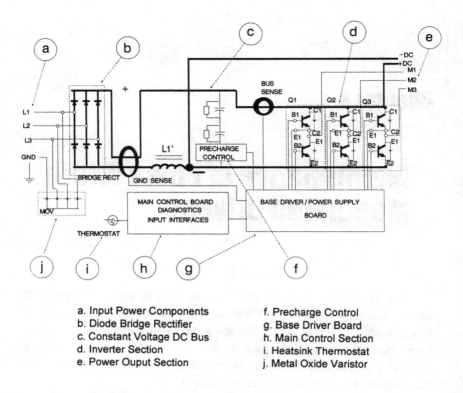

a. Input Power Components
b. Diode Bridge Rectifier
c. Constant Voltage DC Bus
d. Inverter Section
e. Power Ouput Section
f. Precharge Control
g. Base Driver Board
h. Main Control Section
i. Heatsink Thermostat
j. Metal Oxide Varistor

Figure 8-1 Generic Components of PWM Drive

inductors to combat harmonic problems costs less than isolation transformers. It is also important to follow the recommendations for proper grounding.

Diode Bridge Rectifier (Converter), Schematic Item (b)

The rectifier assembly is used to convert the AC input power to DC. The DC output voltage of this assembly will be about 1.4 times the AC input voltage. The rectifiers may be single diodes, or perhaps "twin-packs," depending upon the amperage rating, and will often be mounted on a relatively inaccessible surface that will serve as a heat-sink. Typically, a Metallic Oxide Varistor (MOV) is provided (j) to provide voltage surge protection for the diodes.

Constant Voltage DC Bus, Schematic Item (c)

On a 460 volt drive, the DC bus voltage will be about 650 volts, so caution is advised when probing around in the drive with power on. An indicating lamp will usually show that the bus is hot whenever the voltage is above about 50 volts.

The voltage is stabilized by a bank of capacitors. Since a discharged capacitor appears as a short circuit to the diode bridge, some means of controlling the charging rate is required to protect the rectifiers. Protection is usually provided by a series resistor, with a delayed by-pass contactor, or a solid state current control (f) in series with the capacitor bank. The power factor (Pf) reflected to the line from a PWM drive is constant. However, it *may not be high* unless a DC Bus Inductor (L1') is installed. Small, competitive priced drives usually do not include L1', and their Pf may be constant as low as 70%. A proper inductor can provide constant Pf at 95% or higher under all operating conditions. Sensors to detect grounds and current values are typically included in the DC bus hardware.

Inverter Section, Schematic Item (d)

The power components used in the inverter will vary depending upon the manufacturer, and the output current requirements. Lower horsepower units will typically use power transistors. Higher horsepower units will often use Gate Turn Off (GTO) devices which are a special controllable version of an SCR.

The power components are mounted on a heat-sink, and may be individual components, or constructed as twin-packs. They are usually quite inaccessible, without removing other components. This section changes the DC bus power to variable voltage and frequency.

Power output section, Schematic Item (e)

Terminals M1, M2, & M3 connect directly to the AC motor terminals. Initial direction of rotation must be established between the inverter and the motor. Rotation cannot be changed by changing connections at the drive input terminals. DC bus terminals are also provided. Optional Dynamic Braking units are connected to these terminals. Dynamic brake modules are designed to sense excess DC bus voltage, usually caused by over-hauling loads. If bus voltage goes above a certain threshold, a transistor chopper circuit bleeds the excess voltage across a resistor bank.

Precharge control, Schematic Item (f)

Limits the inrush current to capacitors during power-up cycle.

Base Driver Board, Schematic Item (g)

The base driver board controls the output, using a sequence switching algorithm, to provide a pseudo-sine wave output with proper phase rotation, frequency, and voltage.

Main Control Section, Schematic Item (h)

The main control section usually consists of a main board (motherboard), and one or more auxiliary boards to provide optional features and interfaces. Diagnostic LED's and readouts will be a part of this section. Faults are detected in this area, and acted upon as necessary for reporting status, controlled shut-downs, and panic stops.

Heat-sink Thermostat, Schematic Item (i)

This is an example of one of the protective sensors.

GENERIC FAULTS AND PROCEDURES

You have been the chief maintenance electrician for the Widget Corporation for many years. Whenever a new machine was purchased, you immediately became responsible for its successful installation, performance and maintenance. Using the supplier's schematics and wiring diagrams you have followed the terminal to terminal wiring instructions, and with a little "tweaking" have put each machine on line.

Over time, you become familiar with various idiosyncrasies, make adjustments to compensate for wear and replace control components as they fail. Your standard squirrel-cage induction motors have run for years without any serious maintenance problems.

One day your Production Superintendent announces that the production line is going to be up-dated to operate at variable speeds to optimize output, and make it more flexible. Engineers have decided to purchase brand XYZ's variable frequency drives, (VFDs), and you are to install them on designated motors.

Following their instructions, you make the installation, adjust a few parameters per the manufacturer's installation manual and put the system on line. For the first few days, you are kept busy tuning the system for optimum performance under conditions of varying loads and speeds.

Then, for no apparent reason, one motor starts tripping off intermittently. While running slowly, at normal load, another of your faithful old motors burns out. A third motor, operating at high speed, stalls out as the load is applied. A circuit breaker trips off occasionally for no apparent reason. An electronic timer causes a jam-up because it is tripping too soon. You go through your usual trouble-shooting routine and find that there is no apparent reason for these problems.

In reality, the engineers have ordered this installation without considering the fact that the induction motors operate differently when run at frequencies other than 60 Hz, and the VFDs are tripping off intermittently in a "protect" mode occasionally due possibly to momentary regenerative surges. The motor burned out, without tripping an overload relay, pulling normal current—because its internal fan was ineffective for cooling at the selected operating speed. A motor running at an extended speed range stalled because it had very low pull-out torque operating at high frequency. Timer problems could be caused by harmonic voltage distortion that creates multiple "zero-crossings"—these could be caused by routing the wiring too close to power leads and getting a "coupling" effect. The circuit breaker could be tripping because of current harmonics—undetected by normal metering—that superimpose additional harmonic currents on the circuit.

Obviously this is probably an extreme case, but it illustrates the importance of understanding the operating capability of the motors on variable frequency, and some of the possible pitfalls when adding a large electronic power load to your system.

The following charts, "Generic Faults and Procedures," list many of the typical faults that can occur with drives from any manufacturer. Bracketed letters, e.g. [a], refer to items marked in Figure 8-1. Blank charts are included to allow addition of unlisted faults you may experience. You will find that most portable ammeters cannot respond accurately to the inverter output current because of its carrier frequency and wave-form. A volt-ohmmeter is the only testing device that you will need to follow the procedures listed. If the drive has a current read-out, it will be accurate enough for most practical purposes.

Generic faults and procedures

Fault	Symptom	Probable Cause	Trouble Shooting
	No Control No LEDs ON	No incoming power Blown fuse, Open switch or circuit breaker. [a]	Test for rated voltage at input terminals. Test fuses.
	Blown power fuses	Shorted MOV. [j]	Isolate and check with ohmmeter. Should read max. resistance.
		Shorted diode in bridge rectifier. [b]	Isolate and test with ohmmeter. Reverse polarity across each diode. Should read ohms one direction and pass current with opposite polarity.
		Precharge controls may have failed or shorted. Capacitor inrush too great.[f]	Check components, replace as necessary.
		Shorted capacitors. [f]	Look for ruptured capacitor. Test with ohmmeter for charge and discharge action.

Generic Faults and Procedures

Generic faults and procedures

Fault	Symptom	Probable Cause	Trouble Shooting
Auxiliary fault.	System energized, no response to control. Fault indicated.	Motor thermostat tripped, external safety interlock open.	Allow motor to cool. Correct problem, reset.
Input power loss.	System energized, no response to control. Fault indicated.	Power outage exceeded ride-through time-out [a]	Reset and restart system.
Bus undervoltage.	System energized, No response to control. Fault indicated.	Line voltage dip below minimum required. [a]	Test input voltage. Check for a large starting load that may have pulled voltage down long enough to trip drive off.
Bus overvoltage.	System energized, No response to control. Fault indicated.	Over-hauling load on motor. [e]	Monitor motor current. Add Dynamic Braking kit if necessary. Extend decel time.
		High line voltage. Possibly result of dropping load while PF capacitors are on line.	Test line voltage. If caused by overhauling load, extend decel or add Dynamic Braking.

Generic Faults and Procedures (cont.)

Generic faults and procedures

Fault	Symptom	Probable Cause	Trouble Shooting
Motor stalled.	System energized, no response to control. Fault indicated	Excessive load caused by a continuous overload. [e]	Reduce load, increase gear ratio. DC boost may be needed.
	Motor will not decel on command.	Bus voltage rise. [c]	Bus voltage rise may be clamping decel control. Increase decel time, disable decel clamping, add Dynamic Brake.
Motor overload.	System energized, No response to control. Fault indicated.	Overload has exceeded time limit [e]	Overload parameter may be set too low. Reduce sustained OL with accel and decel adjustments if possible. Increase gear ratio.
Over temperature.	System energized, No response to control. Fault indicated.	Heat-sink temperature is too high. [i]	Heat-sink blocked or dirty. Fan failure. Ambient temperature too high for required load level.
Over current.	System energized, No response to control. Fault indicated.	Current exceeded 180% of max. [e] (Varies with Mfgr.)	Check for short circuit in output wiring, damaged motor, overloaded or stalled motor.

Generic Faults and Procedures (cont.)

Maintenance of Pulse Width Modulated Drives

Generic faults and procedures

Fault	Symptom	Probable Cause	Trouble Shooting
Ground fault.	System energized, no response to control. Fault indicated	Current path to earth ground detected at one or more output terminals. [e]	Check motor and power wiring for any grounded condition. Motor insulation failure. May require output inductors to protect motor from IGBT voltage spikes.
Output short.	System energized, No response to control. Fault indicated.	Short circuit between output terminals. [e]	Check motor and wiring for short circuit. May require output inductors to protect motor from IGBT voltage spikes.
Transistor short.	System energized, No response to control. Fault indicated.	Shorted transistor has been detected. [d]	Check upper and lower portions of each transistor per mfgr's instructions. Replace as required. Use mounting compound per specifications.
Pre-charge open.	System energized, No response to control. Fault indicated.	Pre-charge circuit defective. [f]	Inspect with power OFF. Replace damaged parts.

Generic Faults and Procedures (cont.)

Generic faults and procedures

Fault	Symptom	Probable Cause	Trouble Shooting
Base Driver/ Power Supply	System energized, No response to control. Fault may or may not be indicated, depending upon type of failure.	Possible line voltage spike. Cascading failure from other circuitry connected to main board. [h]	Look at all fuses. Test for input voltage. Substitute a known good board, or boards. Be certain any interface auxiliary boards are properly seated, and connected.
		Cable failure. [h]	Reseat all inter-connecting cables and try to reset.
Auxillary Fault Overload Overcurrent Over-temperature	Motor has difficulty starting load.	Insufficient starting torque.	Gradually increase DC Boost (or IRComp) until motor responds. Excess DC Boost may cause motor overheating. Increase gear ratio to reduce current requirements.
	Frequent motor failure.	Operating too slow with constant torque load.	Increase gear ratio so that motor runs closer to base speed under most operating conditions. (Internal fan is ineffective).

Generic Faults and Procedures (cont.)

Generic faults and procedures

Fault	Symptom	Probable Cause	Trouble Shooting
Auxillary Fault Overload Overcurrent Over temperature Output short	Frequent motor failure.	Overheating	Increase drive HP or add external cooling fan to motor.
		Motor does not have adequate winding insulation.	Rewind motor with higher quality insulation, or replace with a motor rated for inverter service (IGBT drives require 1600 to 2000 volt insulation against the high voltage spikes).
	ADD YOUR OWN NOTES BELOW		

Generic Faults and Procedures (cont.)

Generic faults and procedures			
Fault	Symptom	Probable Cause	Trouble Shooting

Generic Faults and Procedures (cont.)

Generic faults and procedures

Fault	Symptom	Probable Cause	Trouble Shooting

Generic Faults and Procedures (cont.)

Chapter 9

Basics of Applying Drives

One of the biggest source of problems, in working with drive systems, is the improper application of the various drive configurations. AC and DC drives have unique characteristics which must be considered when making your choice. Not only are there differences in torque characteristics, but also considerable differences in costs, line disturbances, power-factor, motor packaging, etc. The size of the drive must be determined on the basis of the *torque* required to do the job. The horsepower can be accurately determined after the torque and speed requirements are identified.

BASIC RELATIONSHIPS

Torque

As referenced in Chapter 2, torque is the force required to turn the input shaft of the machine. It is usually measured in pound-feet (lbft). We can measure this value very simply by applying a wrench to the input shaft and measuring the pound-feet of force it takes to move it with a full load on the system. Measure the length of the wrench handle in feet and, using a simple spring scale, note the value shown when the shaft "breaks-away." If the spring scale is attached one foot from the shaft center, and shows a pull of 30 pounds, we have 30 lbft of torque. (Figure 9-1.)

This is a force. To determine the horsepower, we must identify the maximum speed that is required with this load.

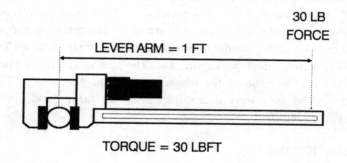

Figure 9-1 Torque

Horsepower

Assume, in our illustration, that this shaft must turn 1750 rpm with the 30 lbft torque. The formula to determine HP is:

HP = torque x Rpm / 5252

HP = 30 lbft x 1750 Rpm / 5252

HP = 10

Basics of Applying Drives

In this illustration we have solved for the friction torque that is always present. If the machine requires high performance, such as fast acceleration and/or deceleration, we must add an additional torque value for the force required to overcome the inertia of the system. This formula is:

$$\text{Torque} = (\text{Inertia} \times \text{change in Rpm}) / (308 \times \text{Time[sec]})$$

Inertia is expressed as Wk^2 or Wr^2 and can be estimated by determining the weight of the rotating mass(lbs), and multiplying by the square of its radius(ft). Wr^2 assumes the mass is concentrated at the maximum radius of the object. Wk^2 assumes that the mass is concentrated at some lesser radius, depending upon the shape of the mass.

For the purpose of illustration, we will consider that we have a 20 lb mass which is concentrated at a 9 inch radius (.75ft).

$$20 \text{ lbs} \times (.75)^2 = Wr^2 \text{ of } 11.25 \text{ lbft}^2.$$

We will assume that this requires acceleration from 0 to 1750 rpm in 3 seconds.

$$Ta = (11.25 \text{ lbft}^2 \times 1750 \text{ rpm}) / (308 \times 3 \text{ sec}) = 21.3 \text{ lbft}$$

Our friction torque was 30 lbft. Add the acceleration torque of 21.3 lbft and we now have a total of 51.3 lbft.

$$HP = (51.3 \times 1750) / 5252 = 17.09 \text{ HP}$$

If we can use a longer acceleration time, a 15 HP drive may do the job. If 3 seconds is critical to the operation and frequent, then we would consider a 20 HP drive. If the duty cycle is not too frequent, we may be able to use a 15 HP drive and allow it to operate in an overload condition during the high torque requirement, provided there is reasonable cooling time between cycles. There is no reason that a motor cannot be operated up to 150% overload for brief periods if the duty cycle is not continuous. The smaller motor will have a lower rotor inertia, and probably have better performance than a larger motor with higher rotor inertia.

Gear box allowance

A slow-turning machine shaft will require a gear reduction between the motor and the shaft. The measured torque value can be divided by the gear ratio, and accel torque by the square of the gear ratio, e.g., a 4 to 1 gear ratio will reflect 1/4 of the friction torque, and 1/16 the acceleration torque. If it is a simple gear box with unknown efficiency, divide again by an efficiency factor of about 0.85; if a worm gear, divide by 0.5. If the efficiency of the gear box is known, use that figure to determine power loss through the gears.

Remember that break-away torque may be 150 to 200% of running torque. We have to supply the worst case torque, considering friction, and accel/decel torques, without exceeding the momentary overload capability of the drive.

As you can see, there are some judgments that must be made to find a balance between price, performance, and maintenance costs when considering an application. The above guidelines should be the basis for determining the real needs of the system. As mentioned, above, if a drive is oversized too much, the inertia of the larger motor may be great enough to reduce the actual response. In that case, the short-time overload capacity should be considered.

Estimating Torque

Sometimes it is necessary to determine the actual required torque on an existing machine, with an installed motor, gears, etc. It may be difficult to make a mechanical determination of the required torque.

In our discussions about AC and DC motors, it was pointed out that current is a good reflection of torque. If we can take ampere readings during each condition of the operation, we can extrapolate an approximate torque value.

The readings should include starting current, acceleration current, running current, and any momentary overload conditions. In other words, we need to know worst case loading, and the duration of that load.

If an AC motor is used, note the full load amperes listed on the motor nameplate. A 10 HP, 460v, 1750 rpm motor will probably have a full load rating of about 13 amperes. One can assume, that under normal running conditions, approximately 25% of the nameplate current will be required continuously for magnetic *excitation* and the remainder will produce *torque*. 25% of 13 amperes = 3.25 amps excitation. This leaves 9.75 "torque" amperes to do the work. The motor will develop:

(30 lbft) / (13-3.25amps) = 3.08 lbft per torque ampere.

Basics of Applying Drives

If our ammeter reads 10 amps, the torque producing current is (10 - 3.25) = 6.75 amperes. We can assume that the motor is developing (6.75 torque amps x 3.08 lbft per torque amp) = 20.79 lbft of torque at 60 Hz.

20.79 lbft x 1750 rpm / 5252 = a demand of 6.9 Hp.

Because there is a constant excitation current required in the AC motor, it must always be accounted for when estimating torque. The relationship of total amperes to torque is not linear, as found with DC motors.

A 10 HP, 1750 rpm, 240 v DC motor will have a current rating of about 38 amperes. Since this motor is also rated at 30 lbft torque, it develops 1.26 lbft per ampere with rated field excitation.

We also know that above base speed, the AC and DC motors have different torque characteristics. The torque of a DC motor varies inversely with the speed in the constant horsepower range. In other words, at double speed, the torque will be one-half and horsepower will be constant.

The torque of an AC motor, operating on variable frequency above base speed, varies inversely as the *square* of the speed at rated slip. The AC motor produces one-quarter of rated torque at double base speed, and rated slip. True constant horsepower at rated slip is not available.

REPLACE DC DRIVE WITH AC SYSTEM

As the reliability of the AC inverter systems has become more predictable, it has opened many opportunities to replace older DC systems with equivalent AC drives. We will show an example in the following pages of a procedure that can be used to make the transition, with the confidence that it will work.

Consider Break-away Torque

The DC motor will normally handle up to 200% of its rated torque to start a load. The AC motor, using a matching inverter on constant torque applications, can be expected to develop about 150% starting torque. (An oversize inverter may allow the motor to develop greater torque—at a price.)

Running torque for the AC and DC motors will be the same up to rated, or base speed. Above base speed, the two motors will "part company" as the torque decline differs between the DC and AC motor. The rated base speed torque of any motor can be calculated by the following formula:

$$\text{Torque} = \text{HP} \times 5252 / \text{nameplate rpm}$$

- Above base speed (NP rpm) we calculate the DC torque as follows:

$$\text{Torque} = (\text{rated torque} \times \text{base rpm}) / \text{rpm}$$

or

$$\text{Torque} = (\text{HP} \times 5252) / \text{new actual speed}$$

- Above base speed, we calculate the AC torque and speeds as follows:

$$\text{Slip} = \text{synchronous rpm} - \text{nameplate rpm}$$

[Synchronous rpm at 60 Hz, will be the next speed higher than nameplate speed, equal to 7200 divided by an even number. e.g. If nameplate value is 1750 rpm: 7200 / 4 = 1800 synch rpm.]

$$\text{New Hz} = (\text{new shaft rpm} + \text{slip}) / (\text{synch rpm}) \times 60\text{Hz}$$

[Always use synch rpm to calculate Hz]

$$\text{Torque} = \text{rated torque} \times (60 \text{ Hz} / \text{new Hz})^2$$

New shaft rpm = (new Hz / 60) x (synch rpm) - slip

Torque by Extrapolation

An existing DC motor nameplate reads 10 HP, 240v, 38a, 1750 / 2700 rpm.

$$\text{Torque} = (10 \text{ Hp} \times 5252) / 1750 \text{ rpm} = 30 \text{ lbft}.$$

This is at base speed (1750 rpm), with full field, 240 volts on the armature and drawing 38 amperes. The motor can develop this constant torque at any speed up to 1750 rpm.

$$30 \text{ lbft} / 38 \text{ amps} = 0.79 \text{ lbft per ampere}.$$

Every ampere in the constant torque range, up to 1750 rpm, will produce 0.79 lbft of torque.

Basics of Applying Drives

For this example, we will assume that the application is operating at 2500 rpm, which is in the constant horsepower (field weakened) range.

Torque = (10 HP x 5252) / 2500 rpm = 21 lbft
21 lbft / 38 amps = 0.55 lbft per ampere

Our ammeter reads 24 amperes.

What is the torque?

24 amps x 0.55 = 13.26 lbft torque

What is the actual demand HP?

HP = (13.26 lbft x 2500) / 5252 = 6.3 actual demand HP

We will now try to apply a 10 Hp AC drive to this application. To keep gearing constant, we will assume an AC motor rated 10 HP, 460v, 1750 rpm, 13 amps.

First, determine the frequency needed to produce a shaft speed of 2500 rpm at rated slip. At 60Hz the synchronous speed is 1750 + 50 rpm slip = 1800 rpm. The required rpm is 2500, so we must add the rated slip of 50 rpm to find the new synchronous rpm. In this case we will look for 2550 rpm, synchronous.

New Hz = (2550 / 1800) x 60Hz = 85Hz.

What is the torque at this frequency? We already know from previous calculations that the motor will be rated at 30 lbft torque in the base speed range.

New torque at 2500 rpm = 30 lbft x $(60 / 85)^2$ = 14.7 lbft (49% of rated torque). We needed 13.6 lbft, so apparently this 10 Hp AC drive could handle the application if starting torque is adequate.

If this same DC motor were drawing 38 amperes at 2500 rpm, we would calculate a torque requirement of 21 lbft to do the job. In this instance, there is no way the 10 Hp AC drive can handle the load at rated slip. We need to stay within rated slip parameters to have good speed regulation.

Consider using a 15 Hp AC drive. The 15 Hp motor will be rated at 45 lbft at full load. In the previous example, we determined that at 85 Hz, there will be 49% torque. 49% of 45 lbft = 22.05 lbft available.

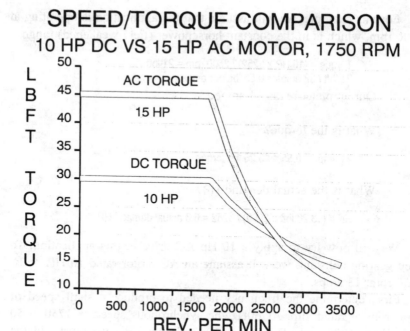

Figure 9-2 AC/DC Torque Comparison

Since this exceeds the 21 lbft required, the 15 Hp system would be correct for operations at this speed and load.

If the starting torque is 200%, or 60 lbft for the DC drive, the AC motor, with 150% of 45 lbft can provide 67.5 lbft, and meet that requirement.

Fig. 9-2 illustrates the torque crossover point for the 10 Hp DC motor and 15 Hp AC motor discussed above. From these discussions, one can see that if we have the nameplate data from the motors, and can take reasonably accurate ampere and rpm readings, we can calculate any conversion that might be required.

It is interesting to note that frame sizes listed on the nameplates can tell you whether your AC motor will fit where the DC is removed. If you divide the first two digits of any NEMA, three-digit frame size by four, it will give you the shaft center-line to base dimension [D] in inches. e.g., Frame 324 will have an 8" center-line to base dimension. This bit of information is useful for maintenance people who might have to find a temporary replacement motor—just check the original nameplate for frame size, and pick any frame equal or less for a spare, and it will fit between the base and center-line. The shaft may or may not be the same diameter. In some cases a 4-pole motor might be temporarily replaced by a 6-pole motor rated at a lower Hp with equivalent torque.

e.g. [A 10 Hp,1750 rpm motor has 30 lbft of torque, a 7.5 Hp,1150 rpm motor has 34.25 lbft of torque]

In summary, there are other factors that must also be taken into consideration, such as regenerative loads, etc. that might not be compatible with a standard AC Drive control package. In some cases, such as hoisting applications, a special regenerative section, or vector-control system would be required to match the energy absorption capability of the DC drive. See Chapter 13 for a case history of an actual project.

Chapter 10

Application Notes

As discussed previously, one should have a basic knowledge of how motors and drive control systems interact in order to apply drives properly. An understanding of the relationships between torque, horsepower, acceleration, and braking characteristics of an application is also required. In this chapter we will discuss how to assemble a drive for a new application. You might be the buyer, or the seller, but both must take a systematic approach to picking the proper equipment.

One cannot ask too many questions. Even a seemingly insignificant bit of information may tip one off to potential pitfalls to be avoided. Asking questions also gives the customer a sense that you are interested in his project, and it gives him an opportunity to explain what the machine has to accomplish. If the customer understands his application, he can ask the proper questions of his supplier to be assured the proper drive is being recommended. Developing a close working relationship with a key person is very important to a successful installation.

DEFINE MACHINE LOADS

Loads will generally fall into one of three categories:

Constant Torque. The load is essentially the same at all speeds. A 10-ton load on a conveyor requires about the same torque at 5 feet per minute as it does at 50 feet per minute. The horsepower demand increases with speed.

Constant Horsepower. The load decreases with increasing speed. This application usually applies to processes that are changing diameter, such lathes, winders, unwinders, etc. With a large diameter, maximum torque and slow speeds are required. As the diameter decreases, torque demand decreases, but speed increases to provide constant surface speed.

Variable Torque. The load increases with speed. This is usually associated with centrifugal fan and pump loads. These applications have the greatest opportunities for energy savings, as the power varies as the cube of the speed. At this writing, many utility companies are offering subsidies to encourage the use of variable speed drives on these applications for energy conservation.

PEAK LOAD

Peak loads are different with various machines. In one case, breakaway torque requirement is very high, as with a sticky conveyor. A high inertia load that requires fast acceleration will likely to have maximum demand during acceleration. Other applications will have maximum demands in the running mode when sudden overloads may appear periodically. It is important to identify the worst-case loading. (See Chapter 14.)

AC OR DC DRIVE?

The choice between AC and DC drives may not be clear at first, although there will usually be a stated preference. Trade-offs must be considered between performance and cost. (See Figure 10-1.)

DEFINE THE MACHINE

We need to define just what the machine does, how it does it. Any interrelationships that are coincidental with the operating cycle must be defined.

Torque

What is the maximum torque required at the input shaft? This can be determined as covered in the "Torque" discussion (Chapter 9), or as provided by the builder of the machine. Are there special high performance requirements for very fast acceleration or deceleration? If so, the inertia and gearing of the system will have to be used to determine *the additional* torque requirements.

AC vs DC Comparisons

	AC	DC
Motor Cost	Low	High
Motor Size	Small	Large
Control Cost	Originally High Getting Lower	Low
Power Factor	Constant (PWM)	Variable
Start Torque (Nominal)	150%	200%
Speed Reg.(Depends)	Slip (3%?)	Tachometer
Motor Inertia	Low	High
Braking (Light) (Hoist)	Dynamic Brake Vector, Regenerative	Dynamic or Regen Brake Regenerative
Constant Hp Range	1.5 to 1 (150% slip)	4 to 1
Isolation Transformer	Optional PWM Required CSI, VVI	Required
Line Disturbances	Harmonics IGBT Harmonics, Notching CSI, VVI	Harmonics, Notching
Motor Availability	Easy	Non-stock
Motor Maintenance	Bearings	Brushes, Bearings

Figure 10-1 AC vs DC Comparisons

$$\text{Friction Torque(lbft)} = \text{Force(lb)} \times \text{Radius(ft)}$$

Reflected friction torque varies inversely with the gear ratio.

$$\text{Accel Torque} = \frac{\text{Inertia}(wk^2) \times \text{change in rpm}}{308 \times \text{time(seconds)}}$$

Reflected accel torque varies inversely as the square of the gear ratio.

Speed

What is the maximum speed required? Is this at full torque or in a lightly loaded condition?

Horsepower. Using the sum of the torques calculated above, and the maximum speed requirement, calculate the horsepower requirement.

$$\text{Horsepower} = \frac{\text{Torque(lbft)} \times \text{Maximum Speed}}{5252}$$

If the highest torque requirements are very short in duration, and within overload capabilities of the drive, it is acceptable to size the drive within those limits. It is important to be aware that too many overloads in a short time may exceed the thermal capacity of the motor and/or its drive.

Sometimes, a small change in gear ratio may optimize the speed/torque needs, and allow use of a smaller drive. These items will become more apparent as we further analyze the machine. The following questions should be considered as a check list.

Machine Starting Issues

- Will the machine operate automatically, or from an operator's control station?
- If from operator's station, what device will be used?
  — Push-button and potentiometer
  — Drum controller
  — Joy-stick (master switch)
- If automatic, what is source of start control?
  — Float switch
  — Pressure switch
  — Clock
  — Thermostat
  — Interlock from another machine
  — PLC, computer, NC, etc.
- What special safety precautions are required for any automatic operations to prevent false starts or damage?
- How often will the machine start?
- What is starting torque?

Application Notes

- What is machine inertia?
- How much time is required for starting?
- Is a smooth start required?
- Under what conditions will the machine start?
  — Loaded
  — Unloaded
  — Varying

Stopping Considerations

- Operator controlled, or automatic?
- Is a quick stop required? (Inertia?)
- Is an accurate stop required?
- Should E-Stop be quick, or coast-to-stop?
- Is a mechanical brake required?

Reversal Considerations

- Does normal operation require periodic reversals?
- How often are these reversals required?
- Is the function basically the same, forward and reverse?
- Is the operation generally in one direction, but requires emergency reversal for maintenance, jams, etc.?

Running Conditions

- Is the machine run at essentially a single constant speed?
- Does it have two or more constant speeds?
- Is variable speed required?
- Over what range?
- Is speed preset or adjustable while running
- Does speed vary through different parts of a cycle?
- Must the speed hold constant with varying load, or can the speed change with load?
- Is there a slow-down required during any part of the cycle?
- Is slow-down abrupt or soft? (inertia, dynamic brake?)

Environment

What are the ambient conditions where the equipment is to be installed?

- Hot or cold (See Chapter 11)

- Wet or dry
- Dusty (conductive ?)
- Corrosive
- High altitude (cooling?)
- Controlled atmosphere, humidity, temperature?
- Size limitations

Motors
- Is an existing motor being used?
- Get entire nameplate information.
- Request resistance and megger readings to determine the condition of the motor.
- Take ammeter readings under various operating conditions to verify suitability for intended operation.

Special Features
- What features do we need for a NEW Motor?
- Horsepower AC or DC
- Base speed.
- Maximum speed if constant horsepower application.
- Enclosure type
- Mounting position, J-box position
  — Vertical
  — Horizontal
  — Feet or flange
    - C or D flange
- Tachometer or other feed-back
- Brake
- Heaters
- Winding thermostat
- Blower
- System Voltage

Protective Features Required
- Overload
- Field loss
- Tachometer loss
- Phase loss
- Overspeed
- Overtemperature, controller and/or motor

Application Notes

- Safety interlocks to other equipment

Monitoring Requirements
- Diagnostics
- Circuit checker
- Load meter, scale?
- Speed meter, scale?
- Ammeter(s)
- Voltmeter(s)
- Fault indicators

Operator Controls
- List those supplied by customer.
- Identify those supplied with drive.
- What signal levels will be supplied for reference?
    — 4 to 10 ma
    — 0 to 5 vdc
    — + and - 10 vdc
    — pulse train
- PLC ?
- Construction
    — Open panel
    — NEMA type

Power Source
- Utility or local source?
- Volts, phase, Hz
- Is isolation transformer or inductor required? If so, supplied by whom?
- Are disconnect switches required?
    — Fused or unfused?
    — Circuit breaker?
- Operator
    — Flange mounted
    — Through-the-door

Loose Items To Include
- Push buttons, lamps, selector switches
- Relays, starters, PLC hardware
- Joy-stick controller(s)
- Potentiometer(s)
- Transducers

SUGGESTED GROUNDING SYSTEM TO MINIMIZE E.M.I. NOISE

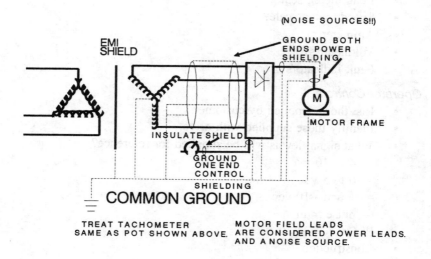

Figure 10-2 EMI Grounding

Wiring Concerns

Electro-magnetic Interference (E.M.I.) is radiated by electronic drive systems. Proper shielding and grounding practices should be followed to minimize and protect sensitive circuits from E.M.I. noise. Figure 10-2 illustrates several ways of controlling EMI. These are valid for both AC and DC drive systems. If cabinet doors and subpanels are not properly bonded they might act as antennas and actually amplify the EMI.

One consideration that is often overlooked is the effect of voltage drop on the *variable voltage power conductors* between the control cabinet and the motor. Excess voltage drop will cause speed regulation problems and torque loss in both DC and AC systems. (Figure 10-3.)

Article 430 of the National Electrical Code (NEC), and Tables 430-149 and 430-150 give instructions for sizing conductors for motor ser-

Application Notes 125

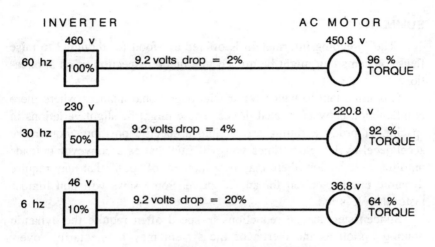

Figure 10-3 Torque vs Voltage Drop

vice. The requirement is that conductor ampacity must be 125% of nameplate current, or 150% of the table values if nameplate data is unavailable. NEC wire tables are designed to prevent conductor overheating, and do not adequately consider torque degradation from line voltage drop. The ampacity ratings are acceptable minimums *if the wire run is not too long.* Present day installations often consolidate drive packages in a common control room, with motors actually installed hundreds of feet away. If the voltage drop at full load and speed exceeds 1%, conductor sizes should be increased beyond the NEC requirements.

Constant torque loads will draw the same current at 10% speed as at 100%. This means that the IR (voltage) drop will be constant. If the voltage drop is 2% at full speed and voltage, it will be 20% at 10% speed. The torque on an AC motor will be reduced to about 64% at this low speed. A DC motor will lose 20% of rated speed, or the tachometer control will have to force excess current (heat?) to compensate for the speed and torque losses.

It never pays to skimp on conductor sizes on any variable voltage power circuit.

SUMMARY

The foregoing information is offered as "food for thought" to raise flags for items that might be missed in a quick inspection of an application.

It is important to watch for special operational nuances where there is interaction between related drives. These might be slight variations in spacing of product, timing arrival into a specific position, momentary accel/decel cycles. Sometimes you will find that as a conveyor unloads onto the next section, there may be a moment of "pull" that may require dynamic braking option for an AC drive. Some saws will pull lumber faster than it is fed.

Sudden step-change reductions in speed often require the dynamic braking option as the inertia of the system may momentarily "overdrive" the motor.

In other words, get a "feel" for how the machine performs. A drive should make the machine operate as designed, not limit the machine to performance of a misapplied drive.

Chapter 11

Heat and Enclosures

Historically, there has not been much concern for sizing enclosures for electrical relays, contactors, etc. For the most part, the amount of heat that was generated could usually be dissipated by any enclosure that could contain them.

With the proliferation of electronic power and control systems, a whole new set of rules have to be considered. Electronic devices consume power and create heat that must be dissipated. All electronic devices have a maximum temperature that can be withstood before they either cease to function, or fail. This maximum ambient temperature is spelled out in the device specifications. Also note that there is a minimum temperature that must be considered which may require the addition of heat, as might be expected in arctic installations.

Frequently, drive modules are field-assembled with special control systems, including PLCs. There is a tendency to make the package as compact as possible by taking physical dimensions and allowing for wireways, but forgetting that these are *heat-producing* devices. There is also a trend of installing all electronics in a common air-conditioned "clean room" which often does not have adequate cooling, especially in hot, high altitude installations.

The author has seen Oregon wood-working plants arbitrarily stop running when the temperature reached 100° F in mountainous areas at an altitude of about 5000 ft. Cabinet doors had to be opened, and compressed air blown through them to get the plant operating again. The enclosures were not only too small, but no one had considered that the lighter air at higher altitudes could not absorb heat as fast as a sea-level installation.

This chapter has been included to bring this often neglected problem into focus. The information may be commonly found in various enclosure manufacturers' catalogue material, and is not claimed to be original in content.

DEFINING HEAT

First, we have to define how much heat we have to contend with. The amount of heat that is generated by the device may be determined in several ways.

Efficiency: In this case, we can look at the power rating of the unit, and take a value based upon its efficiency rating at full load. For example, if it is a 100 kW device rated at 95% efficiency, we would expect the heat losses at full load will be 5 kW.

Watt losses: Watt losses are often specified in manufacturers' instruction manuals, and therefore may not require the above calculations. Add up the watt losses of each device, including control transformers, and use that figure.

Heat generated in BTU/Hr (British Thermal Units per Hour): Heat loss may be given in the specifications as BTU/Hr, which is the most logical value to work with. To convert watts to BTU/Hr, multiply watts by 3.414. In other words, 1 kW becomes 3414 BTU/Hr.

TEMPERATURE CALCULATIONS

It is obvious that if we continually add heat (BTU/Hr) without some means of removing it, temperatures will increase. The maximum temperature rating of the device is usually given in degrees Celsius, e.g., 50°C. In the United States, our ambient temperatures are usually expressed in degrees Fahrenheit, e.g., 68°F. Since BTU is related to Fahrenheit values, it is customary to convert the temperatures to Fahrenheit, to have a common reference system. The following formulas are used for the conversions.

$$\text{Fahrenheit degrees} = ((9/5) \times \text{Celsius}) + 32$$

$$\text{Celsius degrees} = (\text{Fahrenheit} - 32) \times (5/9)$$

HEAT TRANSFER METHODS

Heat transfer depends upon several factors, including the difference between the ambient and allowable maximum temperatures, movement of air, surface area available to radiate heat, and any auxiliary devices that are present to add or subtract from the total heat load.

There are five principal methods that we will examine. These include Enclosure Surface Radiation, Forced Air, Heat Exchangers, Air Conditioners, and Vortex Coolers. In every instance, the requirement is to provide a temperature differential that will facilitate heat transfer at a rate that will prevent the device from reaching its critical temperature. We will now look at each of the above methods of temperature control.

Enclosure surface radiation

Obviously, if we can use the surface to radiate the heat, we may be able to save costs. A problem arises, in some cases, because the enclosure may be oversized beyond acceptable limits. A popular rule-of-thumb for metal enclosures is to allow 1 sqft of area for each 10 watts of heat dissipated. <u>This assumes that the temperature differential is at least 20° C (36° F)</u>. If the temperature differential is equal to or greater than 36° F, this rule may be acceptable.

In calculating the available area for radiation, consider the following limits:

A. The bottom of the enclosure should not be considered in the calculation as heat rises, and has little effect on the bottom surface.

B. The top should be considered at one-half its area if subject to an accumulation of dust and debris. (It is amazing how many rags and old gloves can be found on top of control cabinets, as well as the inches of dust and debris.)

C. The back cannot be considered if the enclosure is to be mounted on a solid wall. It is usable if standing free, or hung on an open rack.

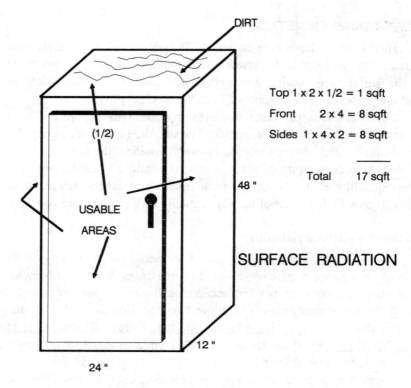

Figure 11-1 NEMA 12 Enclosure

EXAMPLES

Example #1

We will consider a NEMA 12, dusttight enclosure (Figure 11-1) that will be wall-mounted in an area where there will be dust accumulation on its top. Consider a box that is 48" high, 24" wide, and 12" deep.

    Available top area, (12 x 24) / 2 = 144 SqIn / 144 = 1 SqFt.
    Available sides, (48 x 12) x 2 = 1152 SqIn / 144 = 8 SqFt.
    Available front, (48 x 24) x 1 = 1152 SqIn / 144 = 8 SqFt.
    Disregard back and bottom.
    Available Area = 17 SqFt.
    Total heat dissipation using the rule-of-thumb is 10 watts x 17 SqFt = 170 watts.

Heat and Enclosures

Using the rule-of thumb method, a device which is rated at 40° C could be installed in the 48 x 24 x 12 inch box, if the ambient temperature did not exceed 68° F (20° C), and it did not create more than 170 watts of heat.

A more accurate approach is to consider the actual temperature differential. For all calculations, we will assume a standard air density of 0.075 lb/CuFt. At high altitudes where the air density is less, more area would be required to remove equivalent heat. (Derate approximately 2% for each thousand feet above 3300 foot altitude). A heat transfer rate of *1 BTU/Hr per degree F per square foot of area* is the basis of the next example.

Example #2

Using the same size enclosure as above, we will now consider the affect temperature differentials have on the heat transfer capacity. Assume that the equipment has a rating of 40° C (104° F), and the ambient temperature is 50° F. The temperature difference is 54° F.

Our heat transfer is now 1 BTU/Hr x 54° F differential = 54 BTU/Hr per square foot. The total area was 17 square feet, which now calculates to 17 SqFt x 54 BTU/Hr = 918 BTU/Hr. This means that we could handle 918 BTU/Hr / 3.414 = 269 watts. We can actually dissipate 15.8 watts per square foot with this temperature differential..

Example #1 assumed a 68° F ambient temperature which would have given us a 36° F difference. 36°F x 1 BTU/Hr =36 BTU/Hr per square foot. 36 BTU/Hr x 17 square feet = 612 BTU/Hr. This original assumption calculates to be 10.5 watts per square foot of heat removal, which is slightly better than the "rule-of-thumb" method.

The above calculations should be made before considering other means of heat transfer. If our actual load is a device providing 500 watts of heat, the BTU/Hr load will be 500 watts x 3.414 = 1706 BTU/Hr. Since we calculated that our enclosure can only handle 918 BTU/Hr of heat, we need to provide an additional 788 BTU/Hr of cooling from some other source to stay in the same size enclosure.

The installation of a fan in a totally enclosed panel will only move the air and prevent hot spots. It cannot be assumed that an internal fan will remove any additional heat, since the heat is still totally confined in the box, and must rely on radiation from the surfaces to escape.

Forced air

If the installation allows for NEMA 1, drip-proof installation, we can add vents to the enclosure and fans to provide continuous air change. This blows the heat out of the space so it does not accumulate.

It is important that the inlet and outlet vents should be the same size, and arranged so that the air is forced across or through the unit we are trying to cool. It is preferred that the fan be installed at the inlet to pressurize the enclosure slightly, and thus help keep dust and dirt out. The inlet position also produces more turbulence which more effectively picks up heat. This also places the fan motor in the incoming cool air, for longer life expectancy. A plenum ahead of the fan will increase air velocity and make it more efficient. Most enclosure manufacturers provide guide lines for selecting their specific fans (blowers), and their information should be followed for best results.

Most catalogues recommend filters on the ventilation system, sometimes at both intake and outlet vents. Usually, filters are seldom serviced, and soon block the air flow. It is often better to omit filters, and plan to periodically blow dust out of the enclosures. Piping clean air to the fan inlet is the best arrangement.

The standard formula for determining the Cubic Feet per Minute (CFM) follows:

$$CFM = (3160 \times kW) / \text{allowable temperature rise (F)}$$

Our 500 watt heat load, using our box (Figure 11-1) needed to disperse an additional 788 BTU/Hr the from enclosure. 788 BTU/Hr / 3.414 = 231 watts. (0.231 kW.)

$$CFM = (3160 \times 0.231 \text{ kW}) / 36° \text{ F} = 20 \text{ CFM, which requires a very small fan.}$$

We would probably plan to dissipate all the 500 watts with the fan since we are going to the trouble to install one. In this case the calculation would be:

$$CFM = (3160 \times .5KW) / 36° \text{ F} = 44 \text{ CFM}.$$

Pick a fan that will exceed this flow to compensate for pressure drop through the system. The fan will likely be about a 4" assembly, depending upon the manufacturer.

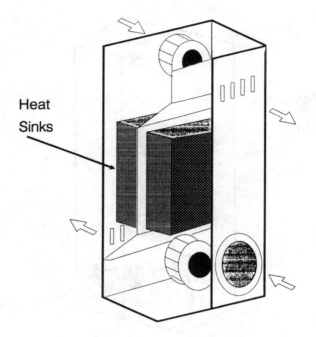

Figure 11-2 Heat Exchanger

Heat exchangers

Heat exchangers (Figure 11-2) are closed loop systems that force enclosure air through a dual plenum, using fan-driven ambient air to remove heat from the system. Since ambient air is used to cool the enclosure, it must always be at a lower temperature than the enclosure. The wider the temperature differential, the more efficiently it will work.

The application of a heat exchanger is fairly straightforward. It requires calculating the panel surface area and the allowable temperature rise, as in the examples #1 and #2.

Beyond that, suppliers will provide a chart to match those values with the unit that can provide the needed additional capacity. This is a good alternative when a NEMA 12 enclosure is required in a compact location. Separate power must be provided for the fans.

Air conditioning

The installation of an air conditioner (Figure 11-3) should be considered after determining that none of the previous methods can do the

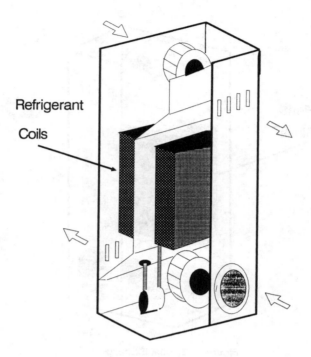

Figure 11-3 Air Conditioner

job. It is important to remember that this is a power-consuming device, and it should be selected as close as practicable to actual demand to prevent excess cycling. If possible, it should be sized to keep the internal temperature about equal to the external ambient temperature to prevent moisture condensation and pooling within the enclosure.

The simplest approach is to follow the guidelines for surface radiation, then select the nearest size air conditioner that is capable of handling the additional cooling load.

If the ambient temperature is *higher* than can be allowed inside the enclosure, then we must calculate the enclosure surface area and the temperature difference, the same as before, except it is for the *heat input* from external sources. Then calculate the BTU load from the internal devices. Add these two loads together to determine total air conditioner (A/C) load.

Heat and Enclosures

Example: Outside temperature 110° F
Inside target temperature 85° F
Internal load 1000w (3414 BTU/Hr)
Surface area of chosen enclosure = 64 SqFt
Temperature differential = 25° F (Ambient higher than internal)
External load 1BTU/Hr x 64 x 25 = 1600 BTU/Hr
Internal load   3414 BTU/Hr
Total   5014 BTU/Hr

A 5000 to 6000 BTU/Hr A/C should be selected.

In an arctic environment, there may be a need to supply both air conditioning and heat, as temperatures may swing from very low sub-zero levels in winter to high temperatures during long summer days. Calculate the heat requirement on basis of enclosure surface area and temperature differential, then convert the BTU/Hr value to watts and apply appropriate heater, with thermostat.

Vortex coolers

Vortex coolers (Figure 11-4) are devices that provide refrigeration using compressed air. The device converts compressed air into two streams of air, one hot and one cold. The cold air is directed into the enclosure, replacing the hot air which is vented outside, along with the hot exhaust of the vortex tube. The device can be used on an air-tight installation and takes up very little space. Hardware typically consists of a fitting with two entrance connections to the cabinet and a filter assembly.

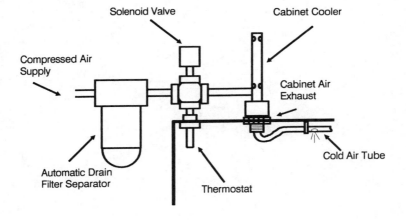

Figure 11-4 Vortex Cooler

A system with 60 psi can provide 2000 BTU of cooling with proper hardware. It does require a source of compressed air that can bear a loss of perhaps 20 to 35 scfm. The air must be as oil-free as possible. A filter must be used to trap water and oil. Thermostatic control is available, with an air solenoid. The devices have relatively low first cost, but power rates will determine the real operating costs to provide the continuous flow of pressurized air.

In this instance, calculate the BTU/Hr load as described in our first illustration, and then pick out the proper unit to supply your need.

Again, the above information is intended to be used as a guide only and is not a replacement for good engineering practice. It is an accumulation of information found in many manufacturers' sales materials and is not represented as an original concept.

Chapter 12

Energy Saving Opportunities Applying Motors and Variable Speed Drives

In recent years, we have seen a high priority placed on the conservation of energy. Environmental concerns have limited the expansion of hydroelectric and fuel burning sources of electrical power. Utility companies have determined that the most economical method of expanding their resources is through more efficient use of the existing facilities, rather than investing in new generating plants.

Conservation is important enough that Bonneville Power Administration and many power utilities have provided funds to subsidize the installation of almost anything that will conserve power. At this writing, in 1996, some of these programs are being curtailed as available funds are being depleted. Much of this "energy" money came from penalties levied against the major oil companies in Alaska, for overpricing their oil and pipe-line charges.

INCREASING MOTOR EFFICIENCIES

An electric motor's price is only a fraction of the cost of the power consumed during a year of operation, so it is very feasible to recover any premium paid for high efficiency with the power savings. There is an industry movement encouraging replacement of standard efficiency motors with those of higher efficiency, sponsored by public energy agencies, such as Washington State Energy Office. See appendix D for information regarding available software they have developed.

The Energy Policy Act of 1992 empowered the Department of Energy (DOE) to establish efficiency standards and testing procedures for most general-purpose, polyphase squirrel-cage motors manufactured in the United States. These new standards apply to continuous-rated, foot-mounted motors, National Electrical Manufacturers Association (NEMA) design A and B, with open and closed enclosures. This includes motors between the sizes of 1 and 200 Horsepower with speeds of 1200, 1800, and 3600 rpm. These requirements become effective in October 1997.

Special purpose motors, such as NEMA C and D polyphase induction motors, synchronous motors, wound-rotor motors, multi-speed motors, vertical motors, and motors manufactured for export are exempted from meeting these efficiency standards.

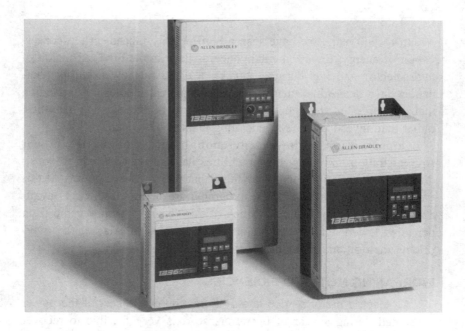

Figure 12-1 Photo, Induction Motor,
Courtesy of Allen-Bradley Co. Inc., Milwaukee, Wisconsin

Many manufacturers have a premium motor line that meets the efficiency requirements, but the majority of the motors do not qualify for these standards. DOE has the authority to extend new standards and testing procedures to fractional horsepower motors if it determines that significant savings can be made, and it can be economically justified.

Figure 12-2 shows the motor efficiency levels prescribed in the Energy Policy Act of 1992, based on nominal full-loads.

DOE Required Motor Efficiencies October 1997						
	Open motors			Closed motors		
Number of Poles	6	4	2	6	4	2
Motor Horsepower						
1	80.0	82.5	-	80.0	82.5	75.5
1.5	84.0	84.0	82.5	85.5	84.0	82.5
2	85.5	84.0	84.0	86.5	84.0	84.0
3	86.5	86.5	84.0	87.5	87.5	85.5
5	87.5	87.5	85.5	87.5	87.5	87.5
7.5	88.5	88.5	87.5	89.5	89.5	88.5
10	90.2	89.5	88.5	89.5	89.5	89.5
15	90.2	91.0	89.5	90.2	91.0	90.2
20	91.0	91.0	90.2	90.2	91.0	90.2
25	91.7	91.7	91.0	91.7	92.4	91.0
30	92.4	92.4	91.0	91.7	92.4	91.0
40	93.0	93.0	91.7	93.0	93.0	91.7
50	93.0	93.0	92.4	93.0	93.0	92.4
60	93.6	93.6	93.0	93.6	93.6	93.0
75	93.6	94.1	93.0	93.6	94.1	93.0
100	94.1	94.1	93.0	94.1	94.5	93.6
125	94.1	94.5	93.6	94.1	94.5	94.5
150	94.5	95.0	93.6	95.0	95.0	94.5
200	94.5	95.0	94.5	95.0	95.0	95.0

Figure 12-2 DOE Required Motor Efficiencies October 1997

OTHER ENERGY SAVING OPPORTUNITIES

Energy saving reflects upon the power cost for any installation. Power costs depend not only on the kilowatts that are used, but also on other factors such as demand and power factor. Large power users can reduce power costs it they take measures to reduce their impact on the utility system's power quality by controlling demand and power factor.

Demand

The electric utility must provide substation capacity to handle the worst case starting loads for a plant. When motors are started across-the-line, there is typically an inrush of about 600% to 800% current. The utility uses a demand meter that determines the maximum power requirement over a fixed period of time, frequently 15 minutes.

The demand charge for the month is determined by this reading and the service equipment has to be sized accordingly. If several large loads are started within the metering time interval, the accumulative result may be very high, even though the average load is relatively light.

The use of "soft-start" controls to accelerate the loads over a period of time will reduce the inrush currents, but may not reduce the demand charges because of the way metering is done. One way to control the demand is to stagger the starting of large loads, so that they do not all occur within the same metering time limit. It is often better to allow a large motor to run through a period of down-time than to make repetitive starts.

Power factor

All induction motors, transformer loads, and older types of flourescent lighting produce inductive reactance which causes the current in a system to lag behind the voltage. (Figure 12-3.) Oversized, lightly loaded motors contribute even greater inductive reactance. Only the "in-phase" current and voltage can be used for real power (kW). The apparent power (kVA) reflects all of the current, including the "out-of phase" current. The reactive component (kVAR) will vary according to the type of load, and if it can be reduced, the power factor will improve.

Briefly, the kVA of a system is the "vector sum" of the kW and kVAR. It is calculated on the basis of the right triangle. kW is the Base, kVAR is the Leg, and kVA is the Hypotenuse. Power factor is the cosine of the angle between the kW and kVA, or kW divided by kVA.

Example:

If the current lags the voltage by 45°, the angle between kVA and kW is 45 degrees. The cosine of 45° is .707 or we can say the kW divided by kVA is .707. In other words, Pf = 70.7%. If real power is 100 kW, the apparent power will be 141 kVA. There is 100 kVAR of

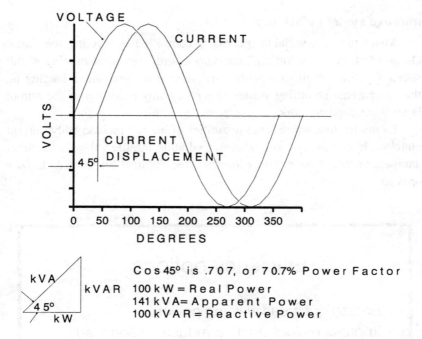

Figure 12-3 Power Factor

inductive reactance in this case, as the 45° angle tells us that the base and leg are of equal value.

At this power factor, the total current in the system is 141% of the effective (real power) current. This extra current produces no work, but heats the transformers and overloads the system conductors. The utility usually charges a stiff penalty for power factor less than 90%.

The power factor can be improved by offsetting the inductive reactance through the addition of power factor correcting capacitors to the system. Capacitive reactance provides a leading power factor which can be used to offset the lagging inductive reactance. Power factor can also be improved by replacing older fluorescent lighting systems with newer high power factor ballasts or fixtures. A good PWM inverter system will give 95%+ Pf at all loads and speeds. DO NOT use capacitor Pf correction with any electronic drive, as a tuned resonance may occur and destroy the electronics and/or the capacitors.

Improve system efficiency

Many times, a variable process is controlled by cycling the motor On and Off, or by "throttling" the output, with the motor running at full speed. Cycling produces repetitive inrush currents and shock loading on the equipment. Throttling wastes energy, as only a portion of the output is producing useful work.

Examples that waste large quantities of energy include eddy current clutches, hydraulic by-pass systems, hydraulic throttle valves, and outlet dampers or inlet vane controls for fans. See Figure 12-4 some *Efficiency options*.

Efficiency options

Search out areas wasting power.
Improve maintenance to reduce friction loads.
Search for sources of excess heat & waste heat.
- Loose electrical connections
- Hot contactors & wiring
- Hot motors

Replace eddy current clutches with drives.
Replace variable pitch belt drives with AC drive.
Use drives to replace dampers and throttle valves.
- Fans & pumps

Select higher efficiency drives even if first cost
 is slightly higher.
Replace 1\2 wave DC systems with full wave.

Consult you power utility energy group.

Figure 12-4 Efficiency Options

Saving energy with AC inverters

By properly applying AC inverters, savings can be made in all three of the above mentioned areas.

Demand

Drives can be effective to reduce demand charges because an application may not have to be brought up to full speed every time it is started. By adjusting the acceleration rate, a load can be brought up to any required speed while applying full torque from zero speed. Across-the-line in-rush currents are eliminated. Demand charges are based upon the total energy used during the metered period.

Power factor

DC drives, current source inverters, and variable voltage input inverters *cannot* be used to improve power factor. All of these drives use silicon controlled rectifiers (SCRs), directly connected to the power source, to convert AC to variable voltage DC power. The output of the SCR is controlled by the "firing angle" of the "gate" control. If one-half voltage is desired, the gate will be turned On half way through the AC pulse. This turns the current On very late in the cycle so that it is out-of-phase with the voltage. The result is that at full voltage, we may attain as high as 92% Pf, but at half voltage it may be in the area of 70% Pf. Power factor varies as the cosine of the firing angle.

Pulse width modulated (PWM) inverters use a diode bridge to provide a constant voltage DC bus, instead of variable voltage. Since there is no firing delay, the PWM drive will give a constant power factor. The lower priced PWM drives may have a constant Pf in the area of 70%. The more sophisticated PWM drive will have an inductor in the DC bus, which, when properly matched with the capacitance of the system, can give constant power factor in the area of 95% or better for all frequencies and loads.

Any load, connected to a *good* PWM drive, is automatically corrected for power factor.

System efficiency

Drives applied to centrifugal fan and pump applications offer the greatest opportunities for energy conservation. Approximately *one-half* of all industrial motors are connected to fans or pumps. These installations are candidates for energy saving if any form of mechanical control is used to modulate the flow. Centrifugal fans, and pumps at zero head, follow the same "affinity laws" for calculating performance. There is an offset to consider when working with head pressures. See Figure 12-5.

By design, every fan and pump has an operating curve that reflects the output volume (flow) as it relates to pressure. (Figure 12-6.) The curve is based upon a specific shaft Rpm.

On the other hand, every system has a curve that reflects the pressure required for a given flow level. Engineers will select a fan or pump with a curve that can produce the maximum requirement of the system curve. Typically, the flow will be controlled by introducing dampers, vanes, or valves in the system.

Affinity Laws for Centrifugal Fans and Pumps

$$Q_2/Q_1 = N_2/N_1$$
Flow is proportional to speed

$$P_2/P_1 = (N_2/N_1)^2$$
Pressure (torque) is proportional to speed^2

$$HP_2/HP_1 = (N_2/N_1)^3$$
Power (kW) is proportional to speed^3

Figure 12-5 Affinity Laws

Energy Saving Opportunities Applying Motors and Variable Speed Drives

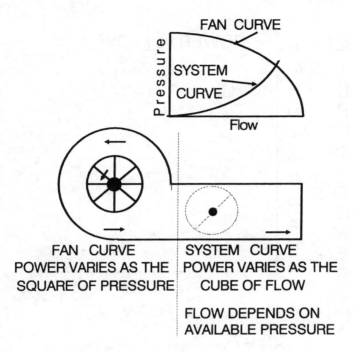

Figure 12-6 Fan Dynamics

Outlet damper (throttle valve)

To reduce the system flow, the outlet damper is partially closed to reduce the available input pressure to the system. At the same time, it is blocking the output of the fan or pump, and creating back-pressure. This back-pressure reduces the volume that can be supplied. The power required varies as the square of the pressure (increasing), and as the cube of the flow (decreasing). The result is that there is very little reduction in power demand as the damper is closed. (Figure 12-7.) Typically, a 20% reduction in flow may not reduce the power requirement by more than about 5%.

Inlet vanes

Inlet vanes basically change the efficiency of the fan. Partially closing the vanes starves the fan so that it can only deliver the air that is available to it. (Figure 12-8.) This is a more efficient method of control than outlet dampers. The power required is almost linear with the flow

Variable Speed Drive Fundamentals

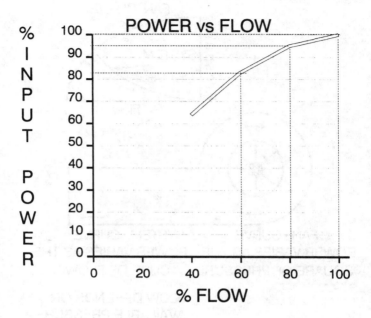

Figure 12-7 Outlet Dampers, Power vs Flow

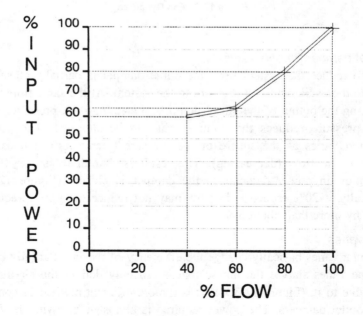

Figure 12-8 Inlet Vanes, Power vs Flow

between 100% and about 60%. Around 60%, most inlet vane systems "go flat" and there is very little change in power as less air is admitted, because of the decreased efficiency of the fan. Reducing the maximum flow by 20% will reflect about an equal reduction in power (20%).

Variable speed control

There is a parallel fan (pump) curve for every increment of speed. In this instance, no obstructions are used in either the intake or output of the fan (pump).

The system flow is dependent upon the pressure available from the source. If one branch of a constant pressure, variable volume system is shut-down, pressure will tend to increase. The pressure can be reduced to normal by driving the fan (pump) at a slower speed to supply the needed flow to remaining branches. There is no significant change in the efficiency of the system, and no additional back-pressure to overcome. This results in power requirements that vary as the *cube* of the flow. (Figure 12-9.) Reducing the flow by 20%, in this case, will reduce the flow demand by almost one-half. An AC meter will show a bit higher kW demand because of the additional constant excitation current an induction motor requires. (Ref. Chapter 8—*"Estimating Torque".*)

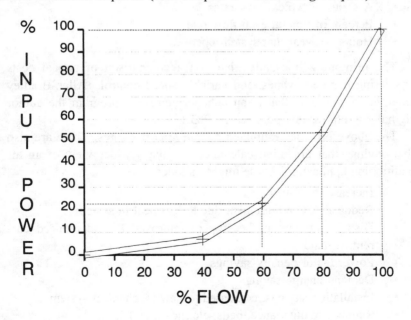

Figure 12-9 Variable Speed, Power vs Flow

At 50% flow:
- Outlet damper will still use about 80% power.
- Inlet vane will require about 60% power.
- Variable speed will require about 13% power.

Existing 3 to 15 PSI transducers can be adapted to directly control fan speeds proportional to pressure.

PAYBACK

The savings can be estimated if one has a record of flow rates required over a repeatable cycle-time for an installation. This cycle might be a day, a week, a month, or whatever, as long as it approximates the real duty cycle of the system. Several manufacturers have free software available for calculating the savings. Typically, you will supply the required variables, as illustrated in Figure 12-10.

- Cost per kWH
- Motor Hp
- Estimated motor efficiency
- Estimated power factor
- A series of typical flow rates
- Percent of time at each flow rate
- Hours per year that system operates

The printout will usually give you a comparison of annual costs considering dampers, vanes, and variable speed control. (Allen-Bradley Co. offers a free disk that is an enhancement of a program the author originally wrote for this purpose in 1983.)

The above provides estimated operating cost figures. There are also other savings that are often calculated to give a wider view of an air conditioning application. These might include:

- Increase in efficiency
- Reduced A/C requirement due to less heat in system
- Distribution savings because of improved Pf and reduced current demand
- Power factor penalty savings
- Demand charge savings
- Installation savings, compared to new mechanical system
- Reduced maintenance, belts, etc.(no-shock)

Energy Saving Opportunities Applying Motors and Variable Speed Drives

ENERGY COST COMPARISON
CONSTANT SPEED vs VARIABLE SPEED FANS

MOTOR HORSEPOWER	100	JOB REFERENCE:	ABC CORPORATION
COST PER KWH	0.04		etc.
ASSUMED MOTOR/FAN EFF	85		etc.
ASSUMED POWER FACTOR	85		
VFD EFFICIENCY	95		
HOURS OF OPERATION	8000	100% KW DEMAND AT THIS EFF	87.76
KVA DEMAND AT THIS EFF/PF	103.25	ASSUMED PERCENT EXCITATION KW	25
COST PER HR AT 100% LOAD	3.51	EXCITATION KW @ 60 HZ	21.94
KVAR AT THIS POWER FACTOR	54.39	TORQUE KW @ 60 HZ	65.82

VARIABLES

TYPICAL POWER REQUIRED FOR OUTLET DAMPERS

FLOW PERCENT	100	90	80	70	60	50	40
PERCENT FAN LOAD	100	98	95	90	83	73	63
EXCITATION KW	21.94	21.94	21.94	21.94	21.94	21.94	21.94
TORQUE KW	65.82	64.64	62.53	59.08	54.63	48.05	41.47
TOTAL KILOWATTS	87.76	86.58	84.47	81.02	76.57	69.99	63.41
COST PER HOUR	3.51	3.46	3.38	3.24	3.06	2.80	2.54

TYPICAL POWER REQUIRED FOR INLET VANES

FLOW PERCENT	100	90	80	70	60	50	40
PERCENT FAN LOAD	100	90	80	71	61	56	53
EXCITATION KW	21.94	21.94	21.94	21.94	21.94	21.94	21.94
TORQUE KW	65.82	59.24	52.66	46.41	40.15	36.86	35.15
TOTAL KILOWATTS	87.76	81.18	74.60	68.35	62.09	58.80	57.09
COST PER HOUR	3.51	3.25	2.98	2.73	2.48	2.35	2.28

TYPICAL POWER REQUIRED FOR VARIABLE SPEED DRIVE   95 % Eff

FLOW & SPEED PCT	100	90	80	70	60	50	40
PERCENT FAN LOAD	105	77	54	36	23	13	7
EXCITATION KW	21.94	19.75	17.55	15.36	13.16	10.97	8.78
TORQUE KW	69.29	53.17	37.34	25.02	15.75	9.12	4.67
TOTAL KILOWATTS	91.23	72.92	54.90	40.38	28.92	20.09	13.44
COST PER HOUR	3.65	2.92	2.20	1.62	1.16	0.80	0.54

LOAD PROFILE CALCULATIONS   8000 HRS

%FLOW SELECTED	%DEMAND HRS	WEIGHTED KW DAMPERS	WEIGHTED KW INLET VANES	WEIGHTED KW VAR SPEED
100	5	4.39	4.39	4.56
90	10	8.66	8.12	7.29
80	12	10.14	8.95	6.59
70	20	16.20	13.67	8.08
60	25	19.14	15.52	7.23
50	18	12.60	10.58	3.62
40	10	6.34	5.71	1.34

CHECK SUM 100 %   77.47 KW   66.94 KW   38.71 KW
(unused %   0 %)

EST. YEARLY SAVING { VARIABLE SPEED VS DAMPERS $ 12405
THIS PROFILE -------- { VARIABLE SPEED VS INLET VANES $ 9037

Figure 12-10 Energy Cost Comparisons

ENERGY AUDITING

When evaluating an energy saving opportunity or verifying results, it is often difficult to develop acceptable data. There may be very little information available to establish a base-line to work from. The problem is further complicated by the characteristics of the AC induction motor.

Energy auditing requires meaningful data to determine the input kW required for a given system flow, before and after the system is modified. The following discussion concerns fan and pump applications. It is a true statement that the shaft horsepower on a pump or fan varies as the cube of the flow (speed), but direct power readings from an AC system will not reflect this relationship. An AC motor has a constant excitation current (about 25% of nameplate amps) that is always present and added to the torque producing current. This constant excitation offset skews the total current compared to the torque current. (Ref. Chapter 9.)

There are some straightforward formulas that can be used to determine fan/pump shaft Hp, based on pressure and flow. We have also seen, from previous discussions, that we can estimate motor torque using nameplate data and ammeter readings.

It is reasonably easy to get ammeter, speed, and pressure readings. It is more difficult to provide instrumentation to measure flow rates. If we can estimate the shaft horsepower (kW) that is being used, and measure the pressure, it is possible to calculate the apparent flow of a system. Pressure can usually be measured with a pressure gauge in a pumping application, or a monometer for fans.

Most ammeters are unreliable when measuring the *output* of a variable frequency drive, because the carrier frequency and wave shape distort the readings. This means measurements will likely have to be made on the incoming lines, ahead of the VFD.

There are several calculations that are necessary to determine flow from input ammeter and pressure readings.

Note that input has constant volts, constant frequency and variable amperes. The VFD output has variable volts, variable frequency, and variable amperes. (See Figure 12-11.)

Energy Saving Opportunities Applying Motors and Variable Speed Drives

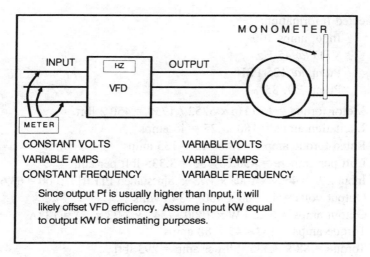

Figure 12-11 Performance Auditing Example

1. Determine motor torque rating using nameplate data.
2. Estimate the excitation amps. [Assume constant 25% of nameplate value]
3. Estimate the torque amps.
4. Calculate rated pound-foot per torque ampere.
5. Determine total kW input from line volts and amps.
6. Determine output volts from speed (hz).
7. Determine output amps using kW and output voltage.
8. Deduct excitation amps from calculated output amps.
9. Remaining torque amps x lbft per torque ampere gives torque.
10. Use torque and speed to determine actual horsepower.
11. Measure pressure.
12. Calculate flow from Hp and pressure.

Example
Motor Nameplate data:
        150 Hp
        1750 rpm
        180 Amps
        460 Volts

Collected Information:
>   Input amps 100a
>   VFD @ 45 Hz
>   Pump @ 125 PSI
>   ·Pump Eff. 85%

1. Motor torque = 150 Hp x 5252 / 1750 = 450.2 lbft.
2. Excitation amps = 180 x .25 = 45 amps
3. Rated torque amps = 180 - 45 = 135 amps
4. Lbft per ampere = 450.2 / 135 = 3.335 lbft per amp
5. Input kW = 460v x 100a x 1.73 x .8[assumed Pf|Eff] / 1000 = 63.6 kW
6. Output volts = 45 / 60 x 460 = 345 volts
7. Output amps = 63.6 kW x 1000 / (345 x 1.73 x .8) = 133A
8. Torque amps = 133 - 45 = 88 amps
9. Torque = 88 x 3.335 lbft per amp = 293 lbft
10. Speed = ((1750+50) x 45 / 60)-50 = 1300 rpm
11. Hp = 293 x 1300 / 5252 = 72.52 Hp
12. GPM = 72.52 x 1713 x .85eff / 125 x 1.0 sp gr = 845 GPM

A fan application can be evaluated in the same manner, except pressure will usually be measured in Inches of Water (PIW). The formula used in step 12, above, to find Cubic Feet per Minute (CFM) is:

CFM = Hp x 6356 x Fan Eff / PIW

USEFUL MOTOR CALCULATIONS

Hp = Torque(lbft) x rpm / 5252   Torque(lbft) = HP x 5252 / rpm

Above are true for AC & DC motors to calculate HP from known torque, and rated torque from nameplate data.

DC Motors

Torque above base speed = Hp x 5252 / actual rpm
Rated lbft per ampere = rated torque / nameplate amps.
Demand torque = amp reading x rated lbft per ampere.
Demand HP = (demand torque x rpm) / 5252.
Speed(full load) = (applied armature volts / rated volts) x nameplate base speed.
Speed above base speed is dependent upon field strength.

AC Motors

Torque above base speed = $(60 / \text{frequency})^2$ x base torque.
(Assume 25% of nameplate amps is excitation amps.)
Nameplate amps - excitation amps = torque amps.
Rated lbft per ampere = rated torque / torque amps.
Demand torque = (ammeter amps - excitation amps) x rated lbft per ampere.
Demand Hp = (demand torque x rpm) / 5252.
Synchronous speed(rpm) = (Frequency x 120) / no. of poles.
Slip = Synch rpm - nameplate rpm.
Frequency = ((New speed + slip) / base synch rpm) x 60.
Shaft speed = ((New frequency / 60) x base synch rpm) - slip

Typical Synch rpms	No. of poles
600	12
720	10
900	8
1200	6
1800	4
3600	2

APPLICATIONS

There are literally hundreds of opportunities to conserve energy using the AC inverter. Following is a brief outline of some real installations that are in service.

- **Fruit warehouse refrigeration control.** (Figure 12-12.) Multiple fans continuously circulate refrigerated air to prevent frost pockets, with a controlled flow rate low enough to minimize product dehydration. No on/off cycling to full speed. Improved Pf. Reduced refrigeration load because fan motors ran cooler. (See chapter 12 for case history.)

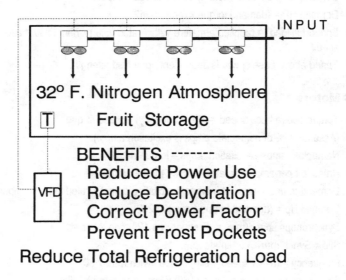

Figure 12-12 Controlled Atmosphere Storage

- **Variable speed refrigeration pump.** (Figure 12-13.) Refrigeration pump was converted to floating head system, allowing the use of variable speed control. On/off cycling is virtually eliminated, with great energy savings. Temperature control of the walk-in freezers improved from 5° differential to 1°. Power rate is about 50 cents per kWH in Alaskan back country villages where these are operating. This installation concept is an "Alaskan Energy Award" winner.

Energy Saving Opportunities Applying Motors and Variable Speed Drives

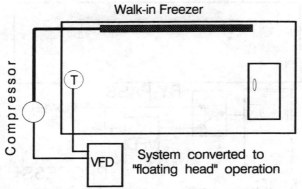

Thermostat controls speed of compressor.
Original cycling held temperature differential to 5° F.
Continuous floating speed range holds to 1° F.

ALASKAN ENERGY AWARD WINNER

Figure 12-13 Variable Speed Refrigeration Pump

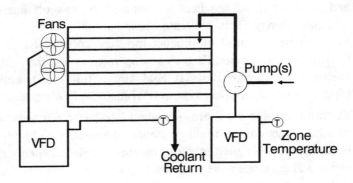

Figure 12-14 Roof Top Chiller

- **Roof top chiller.** (Figure 12-14.) Zone temperature controls pump(s) speed. Return coolant temperature controls fans. The result is a constant temperature differential in controlled space for more efficient operation. No on/off cycling.

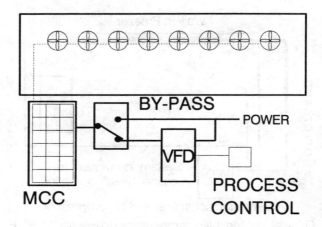

Figure 12-15 Dry Kiln

- **Dry kiln and veneer dryers.** (Figure 12-15.) Motor control center has a variable frequency bus. This controls gangs of fans for optimum air control. After "flashing" surface moisture with high heat and air flow, speed and heat are reduced to draw off intercellular moisture slowly. This prevents "case-hardening" the lumber for more efficient control of product moisture content. Saved steam. Corrected power factor. Fans can be operated at less than full speed during start-up when heavier cold air, at full speed, previously tripped overloads. Bypass system is available for emergencies.
- **Vacuum veneer stackers.** (Figure 12-16.) Use fan speed control to adjust vacuum on multiple zones to match weight of veneer being handled. No need to adjust dampers. Reduced energy requirements and gave operators instant set-up control.
- **Water treatment plants.** (Figure 12-17.) Use level sensors to control pumping rate for continuous control of system. Power factor corrected automatically. Cross-the-line by-pass system is included for emergency operations.

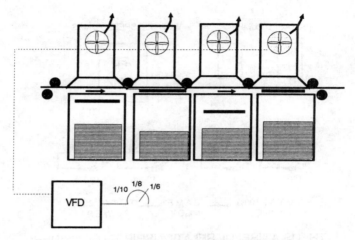

Figure 12-16 Vacuum Veneer Stacker

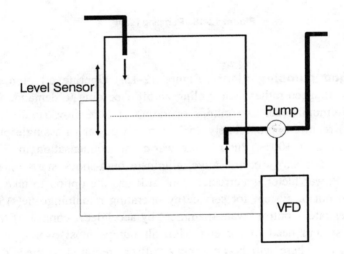

Figure 12-17 Water Treatment Plant

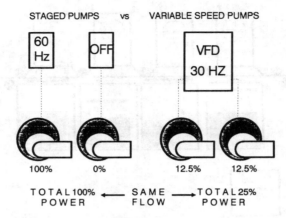

Figure 12-18 Pumping Plant

- **Staged pumping plant.** (Figure 12-18.) Operate all pumps at reduced speed rather than cycling on/off according to demand. With a minimum head, two pumps operating at 50% flow (total 100% flow) require approximately 25% as much power as a single pump operating at 100%. Pumps stay warm (no condensation in windings), seals stay wet and alive, eliminate high-shock starts on system. Power factor is corrected. One still has the option to take any pump out of system for service by operating remaining one(s) at a higher rate. Control "water-hammer" by accel/decel control. If there is a system head to contend with, all pumps must overcome the head, and there will be an energy "offset" required to meet those requirements. The proven savings will more than offset any reduced pump efficiency at lower speeds. (Note that this may take some system engineering to accomplish, but the concept is valid.)

Some applications may require a power by-pass to keep a pump or fan operating in the event of an inverter failure, such as sewer pump, etc. It is important that the *output terminals of the inverter are isolated* from the by-pass line to prevent a back-feed from destroying transistors.

While other motors can be used, the preferred motor for inverter service is a NEMA B, 4-pole motor, with 1.15 service factor and class B or better insulation. Several manufacturers are building inverter-rated motors which meet or exceed these ratings. If applying a drive with IGBT output the motor insulation should be rated not less than 1600 volts.

This brief list of applications may provide some hints that will open doors for many other applications. If you see a controlled pump or a fan, it is likely that it is a candidate for an energy saving opportunity, and your utility may help pay for the installation.

Chapter 13

Case Histories

CASE HISTORY #1 —
A Controlled Atmosphere, Cold Storage Room

Central Washington State is known for its large fruit growing areas, as well as other agricultural crops. There are many controlled atmosphere (CA) cold storage facilities in the area, used to preserve the fruit from the time it is picked until it is sent to market.

The sealed cold storage rooms are held at about 32 degrees F, and have a high nitrogen atmosphere. The nitrogen displaces oxygen to keep the fruit dormant, and prevent aging and decay. The refrigeration system will typically have four or more ceiling mounted refrigeration units with circulating fans. With too much air flow, the fruit tends to dehydrate and shrivel. Too little air movement allows frost pockets to develop. With normal on/off cycling, it is difficult to maintain a correct balance, so there is a certain amount of shrinkage and loss of product that occurs.

In 1987, a young engineer attended one of our seminars where he learned that multiple motors could be operated from a common Variable Frequency Drive (VFD), and power consumption is reduced as the cube of the fan speed. He surmised that he could control air flow and temperature more efficiently by installing a VFD on the cooling fans. By controlling the air velocity, he could reduce the dehydration, while keeping a more even temperature throughout the room. The speed would be controlled within a set band-width, using a temperature sensor and/or other parameters from a PLC controller.

One of the local storage companies was willing to try his idea. The local Public Utility District (PUD) cooperated with him by making a series of system recordings with a Dranetz tester before and after the installation. The tester was connected on the line-side of the VFD.

This room had four, 460v, 5 Hp overhead fans. (See Figure 13-1.) Each fan was an integral part of a refrigeration heat exchanger. The total current draw from the four motors was found to be 33.63 amperes. The plant power factor was 57%. kVA was 28.03. Using 33.63 amperes as the maximum load, he selected a 25 Hp VFD.

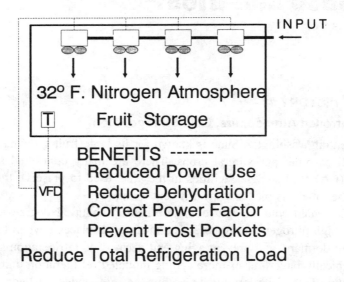

Figure 13-1 Controlled Atmosphere Storage

The motors were all connected to the VFD, and it was set for 60 Hz to get a comparative set of readings. The current was now 21.75 amperes, and the plant power factor was at 91%. One-third of the original current was wattless, because of the low power factor. Input tests were also made at 75%, 50% and 25% speed, with following results.

Hz	Amps	Power factor	kVA
60	21.75	91	17.92
45	12.37	80	9.971
30	6.36	73	5.147
15	3.467	91*	1.986

* Leading power factor appeared on one phase at this low current level, giving the higher power factor reading.

Case Histories

These readings were made *ahead* of the VFD, on the 480 volt line. The VFD was supporting a fair share of the plant power factor, but as the current demand was reduced, the kVA was less, so it could not hold up the rest of the plant load.

The affinity laws for centrifugal fans (and pumps) show that the torque (torque amps) will vary as the square of the speed. (See Figure 12-5.) Also, power (kW) will vary as the cube of the speed. We will now take the readings above, and see how close they came to proving the torque affinity laws.

Since the readings were made ahead of the VFD, the Dranetz tester was looking at constant volts, constant Hz and variable amperes. The motors were operating on variable volts, variable Hz and variable amps.

First, we convert the input to kW for each condition. Ignoring the efficiency for estimating purposes, we then convert the kW to output volts and amperes.

Input kW = V x A x Pf x 1.73 / 1000
Output amps = kW x 1000 / (volts x Pf x 1.73).
Output volts = %Hz x rated volts (480)
Torque Amps = amps - constant excitation amps *
Torque Amps @ %speed = Torque amps x (%speed)^2

* The nameplate record was not provided, but analysis gave us an estimated total excitation current of 9.26 amperes.

% Speed	100	75	50	25
Input kW	16.24	8.003	3.763	1.889
Power factor	91	80	73	91
Output volts	480	360	240	120
Total amps	21.491	16.063	12.415	9.578
Exc. amps (est)	9.26	9.26	9.26	9.26
Torque amps	12.231	6.803	3.155	.75
Calculated torque amps for comparison = 100% torque amps x (% speed)^2				
Sq. rule calc.	12.231	6.880*	3.058	.764
*e.g., 12.231 x (.75)^2 = 6.880				

As you can see from the graph (Figure 13-2), the calculated square rule amperes line very closely matches values derived from the test. The upper line represents the total current that might be read by a meter, and the lower lines represent the torque current required to provide the required brake horsepower. If you calculate the torque for each condi-

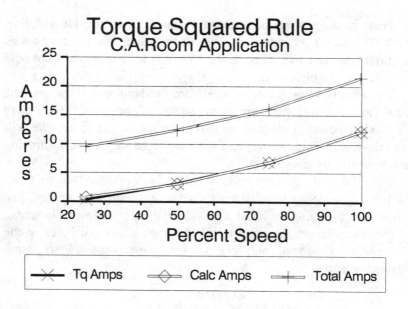

Figure 13-2 Torque, Square Rule Example

tion, and multiply it again by the percent speed, you will get the power curve, varying as the cube of the speed. This will still have a constant excitation kW added to it.

This installation was a success in an area where the power rates are about two cents per kWH. This engineer has completed many of these installations during the past years in spite of the low power rates. The improvement in fruit quality has paid off. He also found that there were other savings—cooler fan motors reduced refrigeration load; improved Pf reduced power losses in the plants; less maintenance as motors are not continually being turned on and off; and improved motor life because windings are at constant temperature and not absorbing moisture.

CASE HISTORY #2 —
Converting DC Driven Pulp Washers to AC Drives

Many paper mills are operating in the forested areas of the United States. Some have been on-line for many years, and have recently been going through modernization as new technology becomes available.

Case Histories

Wood chips are broken down to pulp through chemical processes to provide raw material to make paper. One of the stages in the process involves a washing and bleaching process to condition the pulp for the paper machine.

The washer line carries the pulp on a screen belt through multiple washers. (Figure 13-3.) There are usually at least six machines in the line-up. The screen carries the pulp in and out of washing vats, and over drums at each station. The screen cannot be driven from a single drive sprocket because it must have slack to dip into the washing vats. Therefore, each machine has its own drive connected to the drum. It is important that all drums operate at the same speed to prevent stretching the screen. As a rule, DC motors have been used for the screen drives. A 4-20 ma control loop is usually used to synchronize the motor speeds.

The mill we were looking at had been in operation many years, and the DC motors had had numerous rewinds. The original Silicon

PULP WASHER CONVERSION: DC TO AC

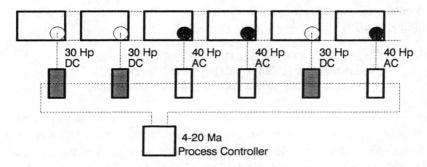

Original motors, 30 Hp, 1750 rpm base speed operating in constant Hp range to about 2300 rpm. 68 LbFt torque available at that speed.
AC motors, 40 Hp, 1750 rpm, driven to 80 Hz provide the 68 LbFt torque at 2300 rpm.

Figure 13-3 Pulp Washer Conversion

Controlled Rectifier (SCR) controls were becoming obsolete, and the enclosures were badly deteriorated from the caustic atmosphere. Since maintenance was becoming more and more prohibitive, the operators considered replacing the DC motors with TENV AC motors and inverter drives whenever an old DC motor had to be removed.

The original motors were rated at 30 Hp, at 1750 rpm. The usual operating speed was about 2330 rpm, which is 130% of base speed. Since the load might vary at times, any replacement AC system had to be able to match the torque of the DC motor at this speed. It also had to be able to provide starting torque up to 200% to match the DC system. Gearing would not be changed, so the AC motor would also have a 1750 rpm base speed, and would have to operate at 80 Hz to provide 130% speed. The following calculations were used to pick the proper AC drive.

> Rated torque of existing DC motor = 30 Hp x 5252 / 1750 rpm = 90 Lbft.
> Torque at 130% speed = 30 Hp x 5252 / (1750 x 1.33)rpm = 68 Lbft.
> Break-away torque = 200% x 90 Lbft = 180 Lbft

A 30 Hp, 1750 rpm, AC motor would have the same running torque (90 Lbft), but typically will only have 150% starting torque (with matching inverter). The torque at 80 Hz could not match the DC motor because torque falls off as a square above base speed with the AC motor. A 40 Hp AC system was selected.

> Rated torque = 40 Hp x 5252 / 1750 = 120 Lbft. (torque to spare)
> Torque at 80 Hz = $(60/80)^2$ x 120 Lbft = 68 Lbft. (equal to DC)
> Break-away torque = 150% x 120 Lbft = 180 Lbft. (equal to DC)

Three AC systems were installed during the first maintenance period. The same 4-20 ma control loop was used to drive both DC and AC drives, without any problems. In the following year or two, the remaining DC drives were phased out. Refer to Figure 13-4 for further details.

The AC system had to be oversized enough for the 80 Hz torque to cross-over at the same point as the DC system at that speed. See Figure 13-5.

Case Histories

DC HP	30		AC HP		40
BASE SPD	1750		SPD 60HZ		1750
FL TORQUE	90		FL TORQUE		120
% SPEED	DC SPEED	DC TORQUE	AC TORQUE	AC SPEED	AC HERTZ
8	146	90	120	100	5
17	292	90	120	250	10
25	438	90	120	400	15
33	583	90	120	550	20
42	729	90	120	700	25
50	875	90	120	850	30
58	1021	90	120	1000	35
67	1167	90	120	1150	40
75	1313	90	120	1300	45
83	1458	90	120	1450	50
92	1604	90	120	1600	55
100	1750	90	120	1750	60
108	1896	83	102	1900	65
117	2042	77	88	2050	70
125	2188	72	77	2200	75
133	2333	68	68	2350	80
142	2479	64	60	2500	85
150	2625	60	53	2650	90
158	2771	57	48	2800	95
167	2917	54	43	2950	100
175	3063	51	39	3100	105
183	3208	49	36	3250	110
192	3354	47	33	3400	115
200	3500	45	30	3550	120

START TORQUE (200%)   START TORQUE (150%)
180 LB FT   180 LB FT

Figure 13-4 Replacing DC Drive with Equivalent AC Drive

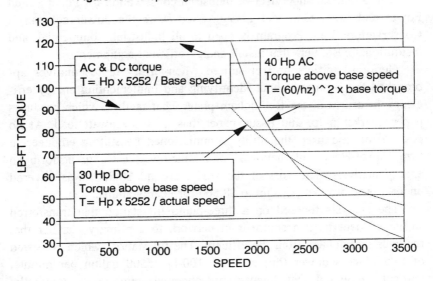

Figure 13-5 AC vs DC Torque

CASE HISTORY #3 —
A Utility-Subsidized Project

(Reprinted with permission from the March 1993 issue of Energy User News, 1 Chilton Way, Radnor, PA 19089.)

Utility Aid Cuts Paper Mill's VSD Payback from 9 Years to 17 Months
By Mike Randazzo

Bellingham, Wash.—Incentive payments and technical assistance provided by the local utility cut the payback of a variable-speed drive installation at a Georgia-Pacific Corp. pulp and paper mill here from nine years to only 17 months, despite unusually low electricity rates in the area.

In addition to a free energy analysis and construction management funding, Georgia-Pacific (G-P) also received $71,132 in rebates for the project from Puget Sound Power and Light Co. The rebate shrank the paper mill's investment in the $84,145 project to just $13,013. Annual utility bill savings are estimated at $9,432, according to electrical engineer Ernie Diaz. He based the projections on a 2.4 cents per kilowatt hour (kwh) energy charge and an annual energy consumption reduction of 393,000 kwh.

Bob Stolarski, supervisor of industrial conservation for Puget Sound Power and Light Co., told *Energy User News* the power company's Conservation Grant Program is open to all industrial, commercial, and institutional customers that install energy-efficient equipment.

Georgia-Pacific's project started when utility representatives approached the company about identifying and implementing energy efficiency measures at the Bellingham plant. That facility, which produces paper, market pulp, and tissue paper, has a 50-megawatt load. At no cost to the end user, the utility commissioned consulting engineering firm Carrol-Hatch International Ltd., Vancouver, B.C., to perform an energy analysis. The cost of the study, around $8,500, was covered under a program that pays up to $8,500 for energy audits.

The analysis focused on a three-pump lift station that transferred sulfite effluent from a sump drain network to a primary clarifier that removes particulates from the effluent. The lift station, which consisted of two 75-horsepower (hp) and one 100-hp, 5500 gallon per minute, vertically mounted centrifugal pump, operated continuously. Normally, the 100-hp pump assumed most of the pumping load and was assisted by a smaller pump when effluent volume exceeded the lead pump's capacity. The third pump was for backup.

According to Charlie Nishi, the engineering firm's vice-president, pump flow is controlled via level transmitters that in turn operate restrictor valves on the discharge end of the pumps. "Considerable" energy losses occurred during low pump output, because the lead pump's motor would turn at the same speed, and use the same amount of energy, regardless of the position of the flow-restricting valves.

To improve the efficiency of the pumping station, Nishi installed an Eclipse Model 3 100-hp variable frequency drive from ICD Drives Inc., a division of Emerson Industrial Control, Grand Island, N.Y., on the lead pump.

By going with variable speed drives the facility was able to substantially decrease the horsepower requirements on reduced flow situations.

In addition to picking up the tab for the energy analysis, the utility also paid for project management. Carrol-Hatch received about $7,500 to assemble equipment contractors and perform initial monitoring. Puget Sound Power and Light Co. puts a $7,500 cap on such "construction management" funding and allows end users to hire engineering consultants of their choice.

The utility is seeing a lot of activity in motors and drives rebates. The reason why Puget Sound Power and Light Co. offers what it likes to call a construction management incentive, is that it finds that many of its industrial clients just don't have the manpower to implement projects like these.

Completed last August [1992], the drive retrofit is the first part of a two-phase project at the lift station. Now that the drive is in place, G-P plans to replace the outdated 100-hp pump and motor assembly with a high-efficiency 10,000-gallon-per-minute submersible version also rated at 100-hp. The new pump and motor will come from Paco Pumps, Inc., Oakland, Calif.

Installing the more efficient pump will not change the plant's load profile much, but will allow the plant to relegate one 75-hp pump to backup duty and remove the third pump to open much-needed space.

The article, above, illustrates one of the many opportunities to work with available utility incentive programs. A number of engineering firms are now specializing in energy conservation projects.

CASE HISTORY #4 —
Irrigation Controls

Many arid areas of the United States are now served by elaborate irrigation systems that require many pumping installations. One of these locations is along the Columbia River between southern Washington and northern Oregon. River water is pumped from the river into a series of feeder canals. In turn, the farms have pumping stations that draw water from the canals and deliver it to large center-pivot irrigating systems.

The huge center-pivot systems often cover up to a quarter section of land when they swing full circle. To reach the squared corners of the field, a trailing swing section is extended with a high pressure "water gun" used to water the odd shaped parcel of land.

A travelling booster pump is turned on whenever the water gun is deployed, which draws down the pressure on the main pivot line. The result is that the main line is starved during this part of the cycle, and there is a visible difference in the crop growth patterns.

By using variable speed pumps for the main center pivot feed, farmers can maintain constant pressure with varying demand. As the water gun is deployed, the pressure sensor automatically increases the pump speed to compensate for the extra volume demanded. This assures that the main pivot line has the constant pressure available to provide proper coverage.

Variable speed pumping can also be used to maintain canal water levels. If a farm shuts down its system, the common feeder canal may overflow if its supply is not controlled. Conversely, as additional irrigation systems are turned on, the canal inflow must be increased to prevent low water levels. Variable speed control is usually more economical, and has less maintenance than staged, on/off pump controls. The first-cost economics of the situation will depend upon whether canal flow rates are relatively constant, or have wide variations.

Variable speed pumps can also be used on deep wells. If there is limited water supply, the pump can be slowed to prevent pumping the well dry. On smaller installations, many AC drives can operate from a single phase source, and deliver three phase output. Since the DC bus does not have full excitation, the inverter will be derated by about 50%. This allows the use of a maintenance-free three phase pump in area where the installation cost of a three phase power line is prohibitive.

CASE HISTORY #5—
Drives Retrofit

(Reprinted with permission from the September 1996 issue of Energy User News, 1 Chilton Way, Radnor, PA 19089.)

Drives Retrofit Saves Plating Facility 451,778 Kwh, 4,429 Million Btu
By Mike Randazzo

Burlington, VT—By outfitting process ventilation fans with variable-frequency drives and digital controls, engineers at Lockheed Martin Armament Systems' metal plating and finishing facility here improved exhaust operation, increased productivity, and lowered energy costs by 29 percent.

The $99,405 retrofit, completed just over a year ago, was a pioneer project in the U.S. Department of Energy (DOE) Motor Challenge industrial showcase demonstration program.

Since the project's completion, Lockheed Martin's annual energy consumption has been lowered by 451,778 kilowatt hours and 4,429 million Btu of natural gas. The combined utility bill savings of $62,282 represent power and fuel usage reductions of 38 and 26 percent, respectively. The retrofit also captured $5,850 in avoided annual maintenance expenses.

The project's anticipated 1.5 year payback should occur within the next few months.

The plating facility is part of a 470,000 square-foot complex that houses the defense contractor's armament systems engineering and administrative offices. Inside, 64 dip tanks and their attendant lateral fume hoods are crucibles for 19 chemically driven processes required for chrome plating, acid etching, anodizing, cleaning, and finishing of metal gun barrels and firing mechanisms. These aluminum and stainless steel components are used to assemble the Gatling-style machine guns deployed on fixed-wing aircraft and shipboard antiaircraft weapons systems.

Added to the main building in 1969, the 20,000 square-foot finishing and plating wing was originally designed to accommodate around-the-clock production schedules manned in three shifts. The addition was equipped with a constant-volume ventilation system for capturing toxic fumes. As part of an ongoing airflow improvement effort from 1991 to 1994, the ventilation system was tweaked and recalibrated to bring it up to current industry airflow standards and provide better air quality. The optimization included a detailed motor survey, air system rebalancing, fume hood geometry modifications, air leakage mitigation, cross draft prevention, and fan speed reselection.

However, in 1994 company officials considered outsourcing the plating operations, a move that prompted a detailed financial analysis of the facility's fixed operating costs, explained Philip Mix, Lockheed Martin's senior facilities engineer.

"We looked at the cost of operation and took into account the fixed costs for chemicals and chemical wastes—costs that are based on our volume of business," he told EUN. "When we looked at all the utilities, we found out that the ventilation in our plating facility was about 74 percent of the cost of running that operation."

Wanting to maintain night-time makeup air temperatures at 60 degrees F and daytime comfort levels at 70 degrees F, Mix observed that he couldn't realize any of the potential utility bill saving with a system that "just pumped air in and out."

Scaled-back demand for product, which dropped the staffing requirement down to one shift, combined with steadily rising energy costs presented the opportunity for Mix to provide automated and energy-efficient controls of makeup and exhaust air fan motors to throttle back fan usage during idle periods.

"We could see our energy costs continuing to climb even at one-third production, and we had no means of handling this fixed operating cost," he related. "It became really obvious that we needed to shut down the volumes of exhaust when these plating tanks were not in use."

The lateral hoods are positioned behind the plating tanks, some of which are heated with steam coils. Air is captured by the hoods and drawn through individual ducts by one of several main branches that are grouped according to chemical compatibility. The air is then discharged to the outside by seven axial, centrifugal exhaust fans ranging from 6,000 to 26,500 cubic feet per minute (cfm). Makeup air is treated by three direct gas-fired rooftop heating and ventilating units. Each of these units is outfitted with a 33,000-cfm centrifugal fan that draws in outside air through filtration elements, moves the air over the gas burners, and supplies diffusers in the plating room's ceiling.

Except for one recently installed unit, these belt-driven ventilation fans were original equipment. The fan motors were supplied by Baldor,

Fort Smith, Ark.; GE Motors, Fort Wayne, Ind.; Marathon Electric Manufacturing Corp., Wausau Wis.; and MagneTek Inc., St. Louis. Motor output varied. Eight of the fans were in the 10-30 horsepower (hp) range and one unit was rated at 100 hp. To maintain fan blower rotations at 800 rpm, motor speeds hovered between 1745-1770 rpm. Supply voltages were 200, 220, 230, and 460 volts AC. Full load current draws ranged from 26 to 113 amperes. Each motor ran at full speed for about 8760 hours per year.

In all, six of the exhaust fans and all three makeup air units were selected for optimization with variable-speed drives. The goals of the project were to regulate exhaust fan speed based on tank temperature and to throttle the makeup air fan's speed to supply slightly less air than the total amount exhausted, maintaining negative pressure, to prevent plating room air from mingling with the air in the main plant.

Each of the nine fan motors were equipped with drives from GE. Line reactors from Rex Manufacturing, Rexdale, Ont., also were installed the drives' electrical feeds to improve power factor and enhance quality.

New tank temperature sensors, air pressure sensors, exhaust flow sensors, safety over-rides, controls hardware, drive control logic, and a graphics package also were added as part of a new Metasys direct digital control energy management system from Johnson Controls inc., Milwaukee. The recently added controls system was part of a plantwide upgrade that replaced a 13-year-old direct digital building automation system from another vendor. Of the drive project's total cost, the end user spent $24,533 to upgrade the plating room controls for variable-speed drive operation. To execute the interface between the drives and the energy management system, about 60 hard data points were added to expand the controls to the plating room ventilation system.

The maintenance savings were a portion of what the facilities department spent to support a service contract taken out on the existing, obsolete plating room energy management hardware. While the project required a rather lengthy debugging period, there were a number of valuable lessons gleaned as the retrofit came to fruition. [See addendum at end of article.] As things now stand, Mix claims that he and his staff are better equipped to match ventilation flows with actual production requirements, while still meeting federal occupational health and safety standards.

"I knew that we cranked back the volume by 40-55 percent," Mix commented. "We brought the ventilation down to what we thought was a safe level and were able to verify that the ventilation was adequate. The main feature of the project for us was that we did not lose any capability with drives, and we got increased flexibility."

Ancillary benefits of the retrofit came in the form of increased troubleshooting ability and more accurate fan control.

The new front-end graphics package enables the end user to have a clearer view of the plating room's ventilation system and to pinpoint malfunctions more accurately. Also, Mix said the new control systems allows him to check airflow more frequently and be more responsive to changing conditions without making costly fan and motor sheave changes. Moreover, the facility is now in a better position to compete with independent plating facilities, Mix told EUN

DOE launched Motor Challenge in 1993 to provide information about efficient motor systems and promote industrial energy awareness through the use of efficient electric motors, drives, driven equipment, and energy-minded motor system integration. Motor Challenge showcase demonstrations high-light industrial end users who deploy efficient motor system strategies that lower energy use and boost productivity. Lockheed Martin Armament Systems was the first company to enlist in the program, which now has 30 partners. In addition to promotional support, DOE funded a post-shakedown independent performance evaluation and provide other technical assistance to the end user.

ADDENDUM
Learning Retrofit Lessons at Lockheed

The following are learning experiences gained by Philip Mix, his staff, and other project partners during the implementation and commissioning processes of Lockheed Martin's drives retrofit. These observations were documented in an independent performance evaluation conducted by Flowcare Engineering Inc., Cambridge, Ont., for the Oak Ridge National Laboratory under the auspices of the U.S. Department of Energy's Motor Challenge industrial showcase demonstration program.

- Current electric rates and equipment duty cycles must be carefully reviewed to ensure the integrity of the savings projections. Initially a blended rate, including both demand and energy charges, was applied to the electricity savings calculations. During the day, however, the fans operate at full power, so demand remained constant.

At the end of 1995 a modification to the site's time-of-use tariff culminated in notable electricity saving for the entire plant. Because of the new lower off-peak rates, the project's annual dollar savings were reduced by about $10,000.

- While they were cheaper than variable-speed drives, two-speed motors were not deemed cost effective in Lockheed's case.

"One of the questions that was raised was, why didn't we go to two-speed motors? We wanted to make sure that we had something on the shelf and that if one of these failed it could be changed out at minimum cost," Mix explained. "We change tank and hood arrangements every so often, and variable-frequency drives allow us to ramp the motors up or down to make temporary adjustments."

- While the energy management system became a vital element of the project as it evolved past its primary function of controlling the drives to become a critical diagnostic tool, the feedback loop from the drives to the building automation system usually reported only motor excitation speed and not motor operation. In order to detect a motor that was locally isolated from the drive or belt failure, the exhaust and air makeup duct-work was equipped with exhaust flow sensors.
- Speed fluctuations within the energy management system parameters, often referred to as "hunting," were mitigated during the project's commissioning.

"We had hunting problems with some of the systems," Mix acknowledged, adding that it is a typical occurrence in this type of retrofit. "When we were checking system pressure we were ramping the speed up and down. Depending on wind conditions and the length of time that doors were being left open, we had to build some dampening into the energy management system so that it wouldn't overreact."

- In order to further reduce noise and increase motor life, line reactors may be required before the motor and the drive.
- Project planners said the vibration caused during low-speed operation may accelerate failure in weak bearings. Of the nine motors that received drives, two underwent bearing replacements to alleviate low-speed vibration problems.

- Project engineers found that in cases where the drives were applied to rewound motors, excessive voltage spikes and noise levels may have caused premature motor failure. According to the performance evaluation, elevated electronic noise and voltage levels may result in motor-line ferroresonance, breakdown of motor insulation, and wire degradation. Engineers recommend replacing rewound motors with new premium-efficiency units, specifying inverter-duty motors or using high-voltage-rated wiring between the drive and the motor.

- Because the full-load operating conditions vary from motor to motor, factory set-points on the drives were modified during the retrofit to match each motor's specific full-load values.

  "When the drives were put in place, we ran them manually," Mix stated. "When you install the drives, you have to calibrate the settings to the actual nameplate information on the motors."

Chapter 14

Conclusion

From the discussions in previous chapters, one will realize that variable speed drives are becoming an important factor in energy economics and industrial automation. It is no longer a subject confined to a "chosen few" specialists. While DC drives will likely continue to be used for many applications, the present-day proliferation of AC drives is going to impact the electrical industry to a much greater extent.

Most electricians (including the author) worked with AC motors for years without any need of understanding how they worked. It was enough to know how to make-up the winding connections and connect three wires—and of course, reverse two of them for proper rotation. When the motor was turned on, it usually ran. If not, just fix the starter controls or replace the motor.

Today's emphasis on efficiency and energy conservation is making it mandatory for most electricians, maintenance staff, and machinery users and builders to learn the basic principles so they can intelligently apply the new AC technology. It is not enough to just substitute a VFD for the motor starter because the ramifications of heat and torque considerations cannot be ignored. A motor operating at rated current and slow speed can burn out because its internal fan is not moving enough air to cool it. Do not use a VFD in place of a "gear-box." A drive should always be set up to work predominately in it's base speed range for best cooling. Overload relays cannot protect it because rated current will not trip it off! A winding thermostat is really the only way to protect a motor in this instance. An external fan may be required, when driving high torque loads at low speeds, to prevent motor overheating.

In chapters 3 and 4 we learned that, using only the nameplate data, one can estimate the available torque for any motor, at any speed, and current demand. In the past it was enough to know maximum available torque at nameplate speed. With this new information, we can now match a motor and drive to any application if the requirements have been determined. We can also use current readings to determine realistic system demands when considering ways to increase efficiency.

Chapter 6, a new chapter in the 2nd edition, discusses the harmonics problems that have become important as the VFD manufacturers have migrated to IGBT technology. As the electronic power loads become predominant, one must be aware of any special installation requirements, to prevent future harmonics problems,

Since the AC drive controls the motor indirectly, it is necessary to understand the causes of the most common faults. It seems that the most frequent nuisance fault is over voltage. It is usually caused by the load characteristics, during deceleration, that tend to over-drive the motor. Acceleration and deceleration rates are much more critical with AC drives than DC systems, so the same rules do not necessarily apply.

As discussed in Chapters 9 and 10, to apply a drive, the worst-case torque must be determined and matched to a motor rated for that torque, or one capable of producing it over a limited overload period. Following are examples of some applications with their probable torque requirements during break-away, acceleration, and peak running conditions. Note that the highest torque appears during different parts of the operating cycle, depending upon the application.

Percent of rated full load torque

Application	Break-away	Acceleration Peak	Running
Agitators			
Liquid	100	100	100
Slurry	150	100	100
Blowers, Centrifugal			
Valve closed	30	50	40
Valve open	40	110	100
Blowers, Positive displacement			
Rotary Bypassed	40	40	100
Card Machines, textile	100	110	100
Centrifuges	40	60	125
Chippers, wood, start empty	50	40	200
Compressor, axial-vane, start loaded	40	100	100
Compressor, reciprocating, start unloaded	100	50	100
Conveyors			
Belt, loaded	150	130	100
Drag (apron)	175	150	100
Screw, loaded	175	100	100
Shaker (vibrating)	150	150	75
Draw press (flywheel)	50	50	200
Drill press	25	50	150
Escalator, start unloaded	50	75	100
Fans, centrifugal, ambient temperature			
Valve closed	25	60	50
Valve open	25	110	100
Fans, centrifugal, hot gases			
Valve closed	25	60	100
Valve open	25	200	175
Fans, propeller, axial-flow	40	110	100
Feeders, loaded belt	100	120	100
Feeders, oscillating drive	150	150	100
Feeders, screw, compacting rolls	150	100	100

Feeders, screw, filter-cake	150	100	100
Feeders, screw, dry	175	100	100
Feeders, vibrating	150	150	100
Frames, textile, spinning	50	125	100
Grinders, metal	25	50	100
Ironers, laundry mangles	50	50	125
Jointers, woodworking	50	125	125
Machines			
Bottling	150	50	100
Automatic buffing	50	75	100
Cinder-block, vibrating	150	150	70
Keyseating	25	50	100
Polishing	50	75	100
Mills			
Flour grinding	50	75	100
Band saw	50	75	200
Mixers			
Chemical	175	75	100
Concrete	40	50	100
Dough	175	125	100
Liquid	100	100	100
Sand, centrifugal	50	100	100
Sand, screw	175	100	100
Slurry	150	125	100
Solids	175	125	175
Planers, woodworking	50	125	150
Presses			
Pellet, (flywheel)	150	75	150
Punch, (flywheel)	150	75	100
Pumps			
Adjustable-blade, vertical	50	40	125
Centrifugal, discharge open	40	100	100
Oil-field, (flywheel)	150	200	200
Oil, lubricating	40	150	150
Oil, fuel	40	150	150
Propeller	40	100	100
Reciprocating,			
positive displacement	175	30	175

Conclusion

Screw-type, primed, discharge open	150	100	100
Slurry handling, discharge open	50	100	100
Turbine, centrifugal, deep-well	50	100	100
Vacuum, paper mill service	60	100	150
Vacuum, others	40	60	100
Vane-type, positive displacement	150	150	175
Rolls			
Crushing, sugar cane	30	50	100
Flaking	30	50	100
Sanders, woodworking, disk or belt	30	50	100
Saws			
Band, metalworking	30	50	100
Circular, metal cut-off	25	50	150
Circular, wood production	50	30	150
Edger	50	40	150
Gang	60	30	150
Screens			
Centrifuges	40	60	125
Vibrating	50	150	70
Separators, Air (fan-type)	40	100	100
Shears, (flywheel type)	50	50	120
Textile machinery	150	100	90
Walkways, mechanized	50	50	100
Washers, laundry	25	75	100

As shown, above, worst-case torque may be demanded in different parts of the operating cycle, so it is important to understand how the systems operate, when the peak loads appear and the duration of those peaks.

In the second edition of this book, we expanded many of our original subjects to give a broader understanding, and have now updated information in the appendices that has evolved since the original writing. We hope that the intent of this book has been met, as we have tried to reduce the mystery surrounding motors and drives. It is possible to have a working knowledge of this subject without being inundated with calculus and quadratic equations. We have not tried to design a motor or drive— just learned how to apply them intelligently.

APPENDIX A
Glossary

Adjustable Speed Drive: A motor, drive controller, (electrical or mechanical) and operating controls that can adjust the output speed of the motor to selected fixed speeds.

Ambient Temperature: Temperature of the air, or surrounding medium, where equipment is operating or stored.

Armature: The rotating member of a DC motor. It may be a permanent magnet type, or the conventional coil-wound assembly with its commutator assembly.

Armature voltage feed-back: A voltage generated in the armature, opposing the impressed voltage, proportional to the speed of rotation. This assumes constant field excitation current. It is used as a negative feed-back to regulate speed when a tachometer feedback is not used. Accuracy is limited to about 2% of set speed.

Band-width: Input frequency range over which the system will respond to a command. Also used to describe a fan or pump speed range where destructive resonance may be present.

Base speed: The manufacturer's nameplate speed where the motor develops rated Hp at rated voltage. A DC motor will reach base speed with rated field current, and maximum rated voltage on the armature. An AC motor will operate at base speed with 60 Hz applied.

Braking: A means of stopping a motor. It may be mechanical, dynamic, or regenerative.

Break-away Torque: Torque required to overcome friction and inertia, to start a machine from standstill. It will be greater than running torque.

Breakdown Torque: Maximum torque an AC motor can develop at rated voltage and frequency.

Bridge Rectifier: A full-wave rectifier assembly used to convert AC current to DC. Current can only flow in one direction through a rectifier element.

Brush: A conductor used to maintain electrical contact between stationary and moving parts of a machine. Usually a carbon compound, mounted in a spring-loaded holder, to conduct current to the commutator of a DC machine. Also used on machines that have slip-rings.

Cemf: Counter-Electromotive Force. The voltage generated in a DC armature in opposition to the applied voltage. Motor speed stabilizes when Cemf approximates the same value as the input voltage.

Closed loop control: A regulator circuit where the reference signal is compared with the actual value of the variable being controlled. A tachometer speed feed-back is a good example. The difference between the reference and feed-back is called an error signal. The polarity of the two signals is in opposition, so the polarity of the error signal is always set up to reduce the error to zero. Regulation depends upon the quality of the feed-back device.

Cogging: A condition often observed during low speed operations with current source and variable voltage input inverters. The motor "steps" between power pulses causing the machine shaft to "jerk". This can be destructive to key-ways, and other mechanical components.

Commutation: The process of interrupting current flow through a controlled rectifier by applying reverse polarity. Also, the coil reversal in a DC armature coil as the brushes switch commutator segments.

Commutator: The insulated, segmented assembly of copper bars, terminated at coil ends on a DC armature. The brush assembly rides on this rotating member and switches current to appropriate coils as the armature rotates. The continuous switching of the coils provides the polarity reversals to provide torque in the motor.

Constant Horsepower Range: The speed range where a DC motor is at rated armature voltage and controlled by field weakening. Since torque falls off as speed increases, constant horsepower is available.

Constant Torque Range: The speed range where the motor can develop constant torque within the cooling limits of the motor. Usually identified as any speed below base speed.

Constant Volts Range: The speed range of an AC motor where frequency is varied while voltage is held constant. Similar to DC constant horsepower, but since torque falls off as a square, true constant horsepower cannot be attained.

Constant Volts per Hertz: (V/Hz) A condition with AC drives where the voltage varies in proportion to the volts. Constant torque characteristics are available as long as this ratio is maintained. Typically applies to speeds below base (nameplate) speed.

Continuous duty: The load level a motor can operate continuously without insulation damage. This depends upon such factors as insulation rating, ambient temperature, and altitude.

Converter: A device for changing AC power to DC. A rectifier.

Current limit: An adjustable protective circuit that sets maximum current that can be drawn by the motor and controller.

CSI: Current Source Inverter. A variable frequency drive using an SCR controlled variable voltage DC bus, and SCR output to control an AC motor. The drive has a large inductor in the DC circuit that is matched to the AC motor. Maximum frequency is about 66 Hz.

Damping: The reduction in amplitude of system oscillation.

Dead band: An input range that can be tolerated without causing a change in the output.

Diode: A device that passes current in one direction, but blocks reverse current.

Dynamic braking: A means of stopping a motor where the energy is dissipated as heat in resistors. With DC motor, the spinning armature is disconnected from the source voltage and connected across a resistor where the regenerative energy is burned off. With AC variable speed controls, the DC bus voltage is monitored, and any excess voltage is discharged to resistors when a maximum threshold is reached.

Duty Cycle: The repeatable operation at differing loads, dependent upon operating and rest times. Duty cycle refers to the temperature rise tolerance of the motor.

Efficiency: The ratio of mechanical output to electrical input expressed in percent. A measure of the energy losses in a system.

Error: The difference between a set-point reference signal and the feedback. An error must be present to make a correction in the system.

Feedback: A signal that reflects the actual operating condition for comparison with the command reference signal.

Field: The stationary part of a DC motor. Usually holds the shunt field coils. The field may have permanent magnets in some motors.

Field Range: The motor speed range from base speed to maximum rated speed, controlled by varying the current in the DC field windings.

Force: The tendency to change the position or motion of an object with a push or pull, usually measured in pounds, ounces, etc.

Form factor: Sometimes called "ripple factor". A comparison of torque and heat producing currents, the AC ripple present in rectified DC circuits.

Four quadrant Operation: A reversing operation where the motor is required to alternately provide torque and absorb regenerative energy. Reversing, high inertia loads require this capability.

Frame Size: A standardized motor dimension system by NEMA. "D" is the dimension from the base to shaft center-line, measured in quarter inches, and shown as the first two digits of the motor frame size. e.g. A 324 frame will have a "D" dimension of 8 inches. This is true with both AC and DC motors.

Full-Load Torque: The torque necessary to produce rated horsepower at rated full-load speed.

Gate: The control element of a Silicon Controlled Rectifier (SCR). A momentary voltage pulse will turn it on. It will not turn off until the source polarity reverses. Also applies to control element of power transistors.

Appendix A Glossary

GTO: A form of SCR that can be turned off before source polarity reverses. Used in high horsepower inverters where transistors cannot be used.

Harmonics: Frequencies that are superimposed on a power system as a result of the non-linear electronic loads. These harmonics distort the voltage and current wave forms and create problems from extra heat, torque loss, and disrupted communications.

Head: A measurement of pressure, usually in feet of water. Used to calculate power required for pump installations.

Horsepower: A measure of work that can be done over time. For motors, Hp = Torque times RPM, divided by a constant 5252.

IGBT: Isolated gate bi-polar transistor. A power transistor capable of very high pulse rates—up to about 20 kHz. It is used at high frequencies, above audible levels, to limit motor noise from the VFD wave form.

Induction Motor: An AC motor where the primary winding (stator) is connected to the power source, and the secondary winding (rotor) carries induced current. There is no electrical connection between the stator and rotor.

Inductor: A ferro-magnetic assembly installed in line and/or load conductors of a VFD to help control the effects of harmonics. Often used in place of isolation transformer. (Does not provide electrical isolation.)

Inertia: A measure of a body's resistance to changes in velocity. Typically expressed as WK^2, which represents its weight times the square of the radius of gyration. It is usually expressed in units of $LbFt^2$.

Intermittent Duty: A motor that never reaches an equilibrium temperature, but is allowed to cool between operations, such crane and hoist operations. Typically rated for 15, 30, or 60 minute duty.

Inverter: A device used to change DC current to AC current. Commonly used in a generic sense to identify an AC variable frequency drive.

IR Compensation: An Armature voltage feed-back used to compensate for speed droop in simple DC drives. As the IR signal increases, a portion is added to the reference signal to compensate for the droop. Regulation limit is about 2% with any stability. On AC drives the signal can be used to increase the input frequency reference to reduce droop.

Isolation Transformer: A transformer that electrically isolates a drive from the AC power line. It protects solid state power components from high transient voltage and current, limits line disturbances created by the solid state devices, and protects the line from grounded drive motors.

Linearity: A measure of how closely a characteristic follows a straight line function. A linear system will give an output in proportion to its input reference signal.

Locked rotor current: The maximum starting current when a motor is started across-the-line. The steady-state current when a rotor is at standstill at rated volts and Hz.

Locked rotor torque: The minimum torque available under locked rotor conditions.

NEC: The National Electrical Code, gives minimum electrical regulations as published by the National Fire Protection Association every 3 years.

NEMA: The National Electrical Manufacturers Association publishes basic uniform standards for electrical equipment. This includes horsepower ratings, speeds, frame sizes, dimensions, etc.

Overload capacity: NEMA specifies that standard industrial motors should have the capacity of withstanding 150% overload for one minute. Special motors may have a different overload percentage and time periods, based on full-load continuous ratings.

Power: Work performed per unit of time. Measured in watts or horsepower: 1 Hp = 746 watts = 33000 lbft per minute.

Power factor: A ratio of real power (kW) to apparent power (kVA) in an AC circuit. The cosine of the angle of displacement between the voltage and current sine waves.

Pull up torque: The minimum torque required to overcome friction, inertia, windage and loading, to accelerate a machine from standstill to full speed.

PWM: Pulse-Width Modulated Inverter drive. The drive uses a diode bridge for a constant voltage DC bus. The DC bus is stabilized by capacitors, and usually will have an inductor to improve power factor. The output is controlled by transistors, or GTO devices and produces variable voltage, variable frequency using a high frequency "chopping" algorithm. The output very closely simulates the line sine wave form.

Reactance: Any force that opposes a change in current flow. Coils produce inductive reactance that delays the flow of current. Capacitors produce capacitive reactance that accelerates the current flow. Reactance is measured in ohms, and affects the current and voltage relationships in AC circuits.

Rectifier: A device that converts alternating current to direct current.

Regeneration: The characteristic of a motor to act as a generator when the load inertia drives the motor above the commanded speed.

Regenerative Braking: A method of retarding or stopping the motor by controlling the regenerated power, and returning it to the power source. This is a method of recovering energy from the system, and improving overall operating efficiency.

Regulation: The ability of a system to hold the commanded speed, expressed as a percent of base or set speed.

Service Factor: A value shown on a motor nameplate indicating a percent of loading above nameplate rating that can be tolerated continuously without serious degradation. A 1.15 service factor indicates that the motor (or generator) can operate at 115% of nameplate value, if necessary, when rated ambient temperature is not exceeded.

SCR: Silicon Controlled Rectifier. A solid state switch, sometimes called a thyristor, that has an anode, cathode and controlling element called a gate. When the gate is momentarily energized, the rectifier will "avalanche" on and continue to conduct until the source voltage reverses. It can be turned on at any point in the AC sine wave, but will only turn off when the sine wave goes through zero.

Slip: The difference in speed between the synchronous (rotating field) speed of an AC motor and the actual rotor speed.

Speed regulation: The percentage change in speed between no load and full load.

Surge protector: MOV (Metallic Oxide Varistor) or special RC networks designed to clip or absorb voltage transients from an incoming AC source.

Synchronous speed: The speed of the stator rotating field in an AC induction motor. Synchronous speed = 120 x Hz / No. of poles per phase.

Tachometer: A small generator usually connected to shaft of AC or DC motor to give a true speed feedback voltage for speed regulation.

Torque: The turning force applied to a shaft. Usually measured in pound-feet or ounce-inches. It is equal to the force applied times the radius through which it acts.

Transducer: A device used to convert one energy form to another. (e.g. mechanical to electrical.)

Transistor: A solid state device having an emitter, base and collector. It can be used as amplifier and for switching control. Power output of many AC drives is controlled by power transistors.

VVI: Variable Voltage Inverter. Similar to CSI variable speed drive in that it uses a variable voltage DC bus, controlled by SCRs. The output bridge circuit switches the variable voltage at the desired frequency. The DC bus is stabilized by capacitors. Output frequency range may be as high as 200 Hz. Multiple motors can be operated from the VVI.

APPENDIX B

Formulas

HORSEPOWER FORMULAS

For Rotating Objects

$$HP = \frac{T \times N}{5252}$$

Where T = Torque (lbft)
N = Speed (rpm)

Objects in Linear Motion

$$HP = \frac{F \times V}{33000}$$

Where F = Force (lbs)
V = Velocity (ft/min)

For Pumps

$$HP = \frac{GPM \times Head \times Specific\ Gravity}{3960 \times Efficiency\ of\ Pump}$$

$$HP = \frac{GPM \times PSI \times Specific\ Gravity}{1713 \times Efficiency\ of\ Pump}$$

Where GPM = Gallons per Minute
Head = Height of Water (ft)
Efficiency of Pump = %/100
PSI = Pounds per SqIn

Specific Gravity of Water = 1.0
1 CuFt per Sec. = 448 GPM
1 PSI = A Head of 2.309 ft (water weight)
62.36 lbs per CuFt at 62°F

For Fans and Blowers

$$HP = \frac{CFM \times PSF}{33000 \times \text{Efficiency of Fan}}$$

$$HP = \frac{CFM \times PIW}{6356 \times \text{Efficiency of Fan}}$$

$$HP = \frac{CFM \times PSI}{229 \times \text{Efficiency of Fan}}$$

Where CFM = Cubic Ft per Minute
PSF = Pounds per SqFt
PIW = Inches of Water Gauge
PSI = Pounds per SqIn
Efficiency of Fan = % / 100

For Conveyors

$$HP(\text{vertical}) = \frac{F \times V}{33000}$$

$$HP(\text{horizontal}) = \frac{F \times V \times \text{Coef. of Friction}}{33000}$$

Where   F = Force (lbs)
V = Velocity (ft/min)

Coef. of Friction
  Ball or Roller Slide = .02
  Dovetail Slide = .20
  Hydrostatic Ways = .01
  Rectangle Ways with Gib = 0.1 to 0.25

TORQUE FORMULAS

$$T = \frac{HP \times 5252}{N}$$

Where   T = Torque (LbFt)
  HP = Horsepower
  N = Speed (Rpm)

$$T = F \times R$$

Where   T = Torque (LbFt)
  F = Force (Lbs)
  R = Radius (Ft)

$$Ta\,(accelerating) = \frac{WK^2 \times Change\ in\ RPM}{308 \times t(Sec)}$$

Where   Ta = Torque (LbFt)
  WK^2 = Inertia at Motor Shaft $(LbFt)^2$
  t = Time to Accelerate (Sec)

Note: To change $LbFt^2$ to $InLbSec^2$, Divide by 2.68,
To change $InLbSec^2$ to $LbFt^2$, Mult. by 2.68.

AC MOTOR FORMULAS

$$Sync\ Speed = \frac{Freq \times 120}{Number\ of\ Poles}$$

Where   Sync Speed = Synchronous Speed (Rpm)
  Freq = Frequency (Hz)

$$\% \text{ Slip} = \frac{(\text{Sync Speed} - \text{FL Speed}) \times 100}{\text{Synch Speed}}$$

Where FL Speed = Full Load Speed (Rpm)
Sync Speed = Synchronous Speed (Rpm)

$$\text{Reflected WK}^2 = \frac{\text{WK}^2 \text{ of Load}}{(\text{Reduction Ratio})^2}$$

ELECTRICAL FORMULAS

Ohms Law

$$I = \frac{E}{R}, \quad R = \frac{E}{I}, \quad E = I \times R$$

Where I = Current (Amperes)
E = EMF (Volts)
R = Resistance (Ohms)

Power in DC Circuits

$$P = I \times E \qquad HP = \frac{I \times E}{746}$$

$$kW = \frac{I \times E}{1000}, \qquad kWH = \frac{I \times E \times \text{Hours}}{1000}$$

Where P = Power (Watts)
I = Current (Amperes)
kW = Kilowatts
WH = Kilowatt Hours

Power in AC Circuits

$$kVA \ (1ph) = \frac{I \times E}{1000} \qquad kVA \ (3ph) = \frac{I \times E \times 1.73}{1000}$$

Appendix B Formulas

Where kVA = Kilovolt Amperes
       I = Current (Amperes)
       E = EMF (Volts)

$$\text{kW (1ph)} = \frac{I \times E \times PF}{1000}$$

$$\text{kW (2ph)} = \frac{I \times E \times PF \times 1.42}{1000}$$

$$\text{kW (3ph)} = \frac{I \times E \times PF \times 1.73}{1000}$$

$$PF = \frac{W}{V \times I} = \frac{kW}{VA}$$

Where kW = Kilowatts
       I = Current (Amperes)
       E = EMF (Volts)
       PF = Power Factor
       W = Watts
       V = Volts
       kVA = Kilovolt Amperes

Calculating Motor Amperes

$$\text{Motor Amperes} = \frac{HP \times 746}{E \times 1.732 \times Eff \times PF}$$

$$\text{Motor Amperes} = \frac{kVA \times 1000}{1.73 \times E}$$

$$\text{Motor Amperes} = \frac{kW \times 1000}{1.73 \times E \times PF}$$

Where HP = Horsepower
       E = EMF (volts)
       Eff = Efficiency (% / 100)
       kVA = Kilovolt Amperes
       kW = Kilowatts
       PF = Power Factor

Calculating AC Motor Locked Rotor Amperes

$$LRA = \frac{HP \times \left(\dfrac{Start\ kVA}{HP}\right) \times 1000}{E \times 1.73}$$

Where LRA = Locked Rotor Amperes
　　　　HP = Horsepower
　　　　VA = Kilovolt Amperes
　　　　HE = Volts

$$LRA\ @\ Freq\ (x) = \frac{60\ Hz\ LRA}{\left(\dfrac{60}{Freq(x)}\right)^{1/2}}$$

Where 60 Hz LRA = Locked Rotor Amperes
　　　Freq (x) = Desired Frequency (Hz)

OTHER FORMULAS

Calculating Accelerating Force for Linear Motion

$$Fa\ (Acceleration) = \frac{W \times Change\ in\ V}{1933 \times t}$$

Where Fa = Force (Lbs) of Acceleration
　　　W = Weight (Lbs)
　　　V = Velocity (Fpm)
　　　t = Time to Accel. Wt. in Seconds

Calculating Min. Accelerating Time of a Drive

$$t = \frac{WK^2 \times N\ change}{308 \times Ta}$$

Where t = Time to Accel. Load in Seconds
　　　WK^2 = Total Inertia (LbFt2)
　　　N = Speed (Rpm)
　　　Ta = Accelerating Torque (LbFt)

Appendix B Formulas

Note: To change LbFt2 to InLbSec2, Divide by 2.68.
To change InLbSec2 to LbFt2, Multiply by 2.68.

$$RPM = \frac{FPM}{.262 \times D}$$

Where RPM = Revolutions per Minute
FPM = Feet per Minute
D = Diameter (Feet)

$$WK^2 \text{ reflected to Motor} = \text{Load } WK^2 \times \left(\frac{\text{Load RPM}}{\text{Motor RPM}}\right)^2$$

Where WK^2 = Inertia $(LbFt)^2$
RPM = Revolutions per minut

Appendix C

The following list is a summary of many companies offering motor and drive products in 1998. The list has been derived from Internet sources, and is only as complete and accurate as those sources. There may be unintentional omissions because of the constant evolution of drive providers, and changing company affiliations. Check with vendor for up-to-date product information.

ABB Drives Inc.
16250 W. Glendale Dr.
New Berlin, WI 53151-2840
(414)785-3200
1-800-752-0696
Fax (414)785-0696

Allen-Bradley/Div of Rockwell International
1201 S. Second St.
Milwaukee, WI 53204
(414)382-2000
Fax (414)382-4444

AC Technology Corp.
660 Douglas St.
Uxbridge, MA 01569
1-800-217-9100
Fax (508)278-7873

Alternative Utility Services of Illinois Inc.
845 Chicago Ave. Suite 2
Evanston, IL 60202
(847)869-8345
Fax (847)869-8341

Amtram
2760 N. Roxbury St.
Orange, NJ 92867-2244
(714)282-0700
1-800-823-7400 (CA only)
Fax (714)283-1300

Baldor Electric Co.
5711 South RS Boreham Jr. St.
Fort Smith, AR 72902-2400
(501) 646-4711
Fax (501)646-5959

Cegelec Automation Inc.
301 Alpha Drive
Pittsburgh, PA 15238
(412)967-0765
Fax (412)-963-3299

Cleveland Motion Controls Inc.
7550 Hub Parkway
Cleveland, OH 44125-5794
(216)524-8800
Fax (216)642-2131

Cole-Parmer Instrument Co.
625 E Bunker Court
Vernon Hills, IL 60061
(847)549-7600
1-800-323-4340
Fax (847)247-2929

Com-trol
2400 Aveberry Court
Conyers, GA 30208
(770)760-1100
Fax (770) 760-1555

Complete Motor Drives Partner, U.S. Dept of Energy
10924 Grant Road, Suite 320
Houston, TX 77070
(281)890-6494
1-888-751-6636
Fax (281)890-6494

Conserve Electric Supply Corp.
3905 Crescent St.
Long Island City, NY 11101
(718)937-6671
Fax (718)937-4057

Control Techniques Drives
8100 W. Florisant Ave.
St. Louis, MO 63136
(716)837-5652
1-800-828-0550
Fax (716)837-5434

Cooper Industries/Crouse-Hinds Div.
PO Box 4999
Syracuse, NY 13221-4999
(315)477-5531
Fax (315)477-5717

Coyote Electronics
4701 Old Denton Road
Fort-Worth, TX 76117
(817)485-3336
1-800-811-3478
Fax (817)485-9437

Custom Controls Co. Inc.
6320 Alder Drive
Houston, TX 77081
(713)666-3258
1-800-231-3112
Fax (713)666-2486

Cutler Hammer Inc.
110 Douglas Road E.
Oldsmar, FL 34677
(813)855-4621
Fax (813)852-6527

Dwyer Instruments Inc.
PO Box 373
Michigan City, IN 46361
(219)879-8000
1-800-872-8141
Fax (219)872-9057

Appendix C

Dynamic Corp.
3122 14th Ave, PO Box 1412
Kenosha, WI 53141-1412
(414)656-4306
1-800-548-2169
Fax (414)656-4328

Energy Controls-Engineering & Environmental Services
607 10th St., Suite 203
Golden, CO 80401
(303)273-9001
Fax (303)277-1522

EnerShop Inc.
1616 Woodall Rogers Freeway
Dallas, TX 75202
(214)777-1619
Fax (214)777-3700

Forest Electric Corp.
2 Penn Plaza
New York, NY 10102
(212)318-1500
Fax (212)318-1789

GE Co./GE Business Information Center
320 Great Oaks Office Park
Albany, NY 12208
(518)869-5555
1-800-626-2004
Fax (518)886-9282

Hitachi America Ltd.
660 White Plains Road
Tarrytown, NY 10591
(914)631-0600
Fax (914)631-3672

Hoffman Controls Corp.
2463 Merrell Road
Dallas, TX 75229-4516
(972)243-7425
Fax (972)247-8674

Humidaire Co. Inc.
2481 Wharton St. PO Box 21756
Philadelphia, PA 19146
(215)467-4646
Fax (215)467-1667

ITT Domestic Pump
8200 N. Austin Ave.
Morton Grove, IL 60053
(847)966-3700
1-800-542-6655
Fax (847)966-9052
Fax (847)965-8379 (ITT Fluid Handling)

Joliet Equipment Corp.
PO Box 114
Joliet, IL 60434
(815)727-6606
1-800-435-9350
Fax (815)727-6626

Leeson Electric Corp.
PO Box 241
Grafton, WI 53024-0241
(414)-377-8810
Fax (414)-377-3440

Liberty Technologies Inc.
Lee Pard 555 North Lane
Conshohocken, PA 19428-2208
(610)834-0330
Fax (610)834-0346

Magnetek Corp. Info Center
26 Century Blvd., PO Box 290159
Nashville, TN 37229-0159
1-800-624-6383
Fax (615)316-5151

Mitsubishi Electronics America Inc.
500 Corporate Woods Parkway
Vernon Hills, IL 60061
(847)478-2100
Fax (847)478-2283

Motortronics
13214 38th St. N.
Clearwater, FL 34622
(813)537-1819
Fax (813)537-1803

NECA Inc.
3 Bethesda Metro Center, Suite 1100
Bethesda, MD 20814
(301)657-3110
1-800-888-6322
Fax (301)215-4500

Plant Engineering Consultants Inc.
521 Airport Road
Chattanooga, TN 37422
(423)892-7654
1-800-437-3311
Fax (423)-894-0497

Robicon
500 Hunt Valley Drive,
New Kensington, PA 15068
(412)339-9500
Fax (412)339-8100

Saftronics Inc.
5580 Enterprise Parkway
Ft. Myers, FL 33905
(914)693-7200
1-800-533-0031
Fax (914)693-2431

Sterling Electric
16752 Armstrong Ave.
Irvine, CA 92714
1-800-654-6220
Fax (714)-474-0543

Stober Drives
1781 Downing Drive
Maysville, KY 41056
(606)759-5090
Fax (606)759-5045

Sumitomo Machinery Corp.
4200 Holland Blvd.
Chesapeake, VA 23323
(757)485-3355
1-800-SM-CYCLO
Fax (757)485-3075

Teco-Westinghouse Motor Co.
PO Box 277
Round Rock, TX 78680-0277
(512)255-4141
1-800-451-8798
Fax (512)244-5512

Thor Technology Corp.
N56 W13605 Silver Spring Drive
Menomonee Falls, WI 53051
(414)252-2200
Fax (414)252-2201

Toshiba International Corp.
13131 West Little York Road
Houston, TX 77041-5807
(713)466-0277
1-800-231-1412
Fax (713)-466-8773

Tower Tech Inc.
PO Box 1838
Chickasha, OK 73023-1838
(405)222-2876
1-800-256-0349
Fax (405)222-2885

TriVar Inc.
2625 Kimbel St. Unit 22
Mississauga, ONT L5S 1A9
(905)671-1744
Fax (905)671-8709

US Energy Controls
3634 Deedham Drive
San Jose, CA 95148-3111
(408)274-5700
Fax (408)238-8330

Unosource Controls Inc.
112 Commerce Blvd.
Loveland, OH 45140-7126
(513)583-4100
Fax (513)583-4104

Warner Electric/Superior Electric
383 Middle St.
Bristol, CT 06010
(860)585-4500
1-800-787-3532
Fax 1-800-821-1369

Warner Technologies Inc.
11859 Wilshire Blvd.
Los Angeles, CA 90025-6601
(310)444-6616
Fax (310)444-0478

Waukesha Engine Div.
1000 W. St. Paul Ave
Waukesha, WI 53188-4999
(414)547-3311
Fax (414)549-2795

York International Corp./Applied Systems Group
PO Box 1592
York, PA 17405-1592
(717)771-7890
1-800-861-1001

Appendix D

<u>WSU Energy Program</u>
1-800-862-2086
http://www.motor.doe.gov
E-mail, motorline@energy.wsu.edu

<u>MOTOR MASTER [MM+]</u>
This comprehensive motor data base, originally produced by Washington State Energy Office, is a highly recommended tool for selecting high efficiency motors. It lists information for almost every new motor that is manufactured, and includes comparisons for calculating costs and energy savings, and in many cases it calculates rebates offered by local utilities.

<u>Paragon Technologies/FESTech Software Solutions</u>
2747 S.O.M. Center Road, Suite D
Willoughby Hills, OH 44094
(440)944-9980, Fax (440)944-9988
Email, info@festech.com

AC Drives Trainer

This Paragon Technologies computer-training course covers AC drives theory and applications. Material is directed toward Allen-Bradley (AB), Bulletin 1336 drives, but is generic enough to be valuable for understanding most PWM drives. A companion study handbook, *Drive Basics, is* available from Control Services (below) as a reference guide [$25].

ShopDoctor™

ShopDoctor is an in-depth, factory floor trouble-shooting tool. Other products include courses for Fanuc Servo drives; A-B PLC-5 and SLC-500; CNC and DNC; and SoftShop Toolbox™ for managing applications on the shop floor computer.

Control Services

812-163rd Ave SE
Bellevue, WA 98008-4958
(425) 746-1653, Fax (425) 957-7949
E-Mail, caphipps@juno.com

Companion study handbook, *"Drive Basics"* for AC Drives Trainer (above).
Wire bound with 168 pages. $25.00 plus S&H

Motor and Drive Templates

This 3-1/2" disc contains ten (10) spreadsheet templates, several with graphs, that can be imported into Quattro Pro, MSWorks, Excel, or 1-2-3 spreadsheet programs. They are written in *.wks format. Many of the motor speed/torque charts and graphs illustrated in *Variable Speed Drive Fundamentals* are derived from this set of templates. The Read.me text describes the contents briefly, and can be read in any "notebook" type text program. *The templates must be opened in a spreadsheet program.* [Cost to purchasers of this text, $10 for programmed disk, shipping prepaid in continental USA.] It can be downloaded, free, from the internet at [http://weOl.energy.wsu.edu/ep/eic/eicfiles.htm]. File name is "HDRIVES.ZIP." It is listed under "popular files" in the Energy Ideas Clearing House.

Appendix D

Files:

ACDC.wks Compare speed/torque data between a DC and an AC motor

ACTORQUE.wks Speed/torque curve for AC motor to 200% speed.

CALCCFM.wks Estimate fan CFM and pump GPM from pressure and amperes.

CARRIAGE.wks Calculate power required for high performance sawmill carriage drive, and the acceleration time for a bandmill.

CLUTCH.wks Estimate energy losses in Eddy Current Clutch.

DCTORQUE.wks Speed/torque curve for DC motor to 2005 speed.

DYNBRAKE.wks Calculate Dynamic Brake wattage and duty cycle.

ENERGY92.wks Energy savings comparing vanes, dampers, VFD

HEAT.wks Estimate enclosure size for given heat load and ambient temperature.

PUMP.wks Determine motor size from pressure and flow. Compare costs of throttle control vs VFD.

Index

A

Air Conditioner 2, 127, 129, 133-135, 148
Amplifiers 19, 20, 21
Applications 5, 6, 15, 22, 23, 28, 54, 55, 66, 70, 80, 86, 107, 110, 111, 113, 115, 117, 118, 122, 126, 133, 143, 148, 150, 152, 154-159, 177-180
Armature 7, 8, 15-21, 23, 26-28, 51, 57, 68, 112
Auditing 150-152, 168

B

Band-width 72, 80, 81, 88, 183
Base speed 18, 19, 55, 86, 111-113, 122, 153, 166, 177, 183
Brake-to-stop 79, 84, 90
Braking 10, 11, 42, 79, 95, 117, 121, 122, 126
Break-away torque 110, 111, 118, 166, 178-180, 184
Breakdown torque 35, 184
British thermal units (BTU) 128, 131, 132, 134-136, 171
Brush 9, 16, 25-29

C

Capacitor 74, 96, 141
Case history 115, 154, 161-176
Celsius 128, 129, 131
Coast-to-stop 22, 79, 84, 121
Cogging 60, 62, 65, 184

Commutator 15, 25, 28, 29, 58, 184
Consequent-pole 44-46
Constant horsepower 18, 45, 55, 111, 113, 118, 122, 185
Constant torque 18, 45, 51, 54, 78, 111, 112, 117, 125, 185
Constant volts 54-57, 60, 63, 69, 143, 150, 163, 185
Convert 9, 49, 57, 94, 128, 135, 143, 154, 163, 164
Counter EMF, Cemf 16, 17, 20, 21, 22, 23
Current 6, 7, 9, 11, 12, 15-23, 28, 29, 31, 34-36, 38, 40, 42, 45, 47-49, 51, 54, 55, 57, 60-63, 66-69, 71-74, 78-81, 83, 86, 88-90, 95-97, 99-103, 110, 111, 125, 140-143, 147, 150, 162, 163, 173, 174, 177, 178, 185
Current amplifier 20, 22
Current source inverter (CSI) 61-63, 185

D

DC injection brake (hold) 79, 84, 90
Demand 3, 5, 12, 40, 48, 66, 68, 111, 113, 117, 134, 139, 140, 143, 145, 147, 148, 153, 158, 163, 170, 172, 174, 178, 180
Diode 24, 63, 64, 94, 95, 143, 185
Drift 65
Drive systems 11, 19, 20, 22, 71, 72, 75, 107, 124, 127, 177, 178, 181
Dynamic brake 10, 22, 23, 79-81, 95, 99, 100, 121, 126, 185

E

Eddy current coupling 6, 7, 142
Efficiency 1, 4, 6-11, 19, 25, 45, 62, 63, 65, 110, 115, 128, 132, 133, 137-139, 142, 143, 145, 147, 148, 155, 156, 158, 161, 163, 168, 169, 172, 174, 176-178, 186
Energy 1, 2, 6, 8, 10-12, 19, 23, 33, 34, 40, 41, 48, 55, 75, 79, 118, 137-159, 168, 169, 171-175, 177
Environment 1, 2, 5, 8, 19, 29, 121, 135, 137
Error signal 20
Excitation 15, 17, 19, 23, 45, 57, 68, 110, 111, 147, 150-153, 163, 164, 170, 175
Exciters 7, 8, 19
Extrapolation 110, 112

F

Fahrenheit 128, 129, 131, 135
Fan 31, 80, 97, 100, 102, 103, 118, 131-133, 142-152, 154-157, 159, 161-164, 171-174, 177, 179, 180
Feed-back signal 19-22, 66, 122, 186
Field 6-8, 15-19, 23, 24, 27, 28, 33, 34, 42, 45, 47, 48, 51, 58, 66, 68, 79, 122, 153, 186
Flow 28, 31, 34, 54, 132, 136, 161, 169, 170, 171, 173-175
Flux 15, 17, 34, 48, 55, 66, 90
Form factor 23, 186
Four quadrant 23, 186

G

Gate turn off (GTO) 60, 187
Gears 102, 109, 110, 113, 118-120, 166, 177

H

Harmonics 71-75, 78, 90, 93, 97, 178, 187
Heat 6-8, 10, 11, 13, 23, 24, 34, 36, 39, 44, 58, 64, 71-74, 97, 99, 100, 102, 103, 122, 125, 127-136, 141, 148, 156, 172, 178, 179
Heat exchanger 129, 133, 162
Heat transfer 129, 131
Heat-sink 94-96, 100
Horsepower (Hp) 55, 56, 95, 107, 108, 111-115, 117, 118, 120, 122, 138, 139, 148, 150-153, 162, 166, 169, 173, 187
Hydraulic drives 3, 5, 142

I

IGBT 64, 65, 75, 101, 103, 159, 187
Inductor 19, 38, 61, 75, 93-95, 101, 123, 143, 187
Inertia 23, 63, 70, 99, 109, 110, 118, 119, 121, 126, 187
Inlet vane 142, 145, 146, 148, 149
Invert 9, 57
Inverter 57, 58, 61, 65, 78-81, 93, 95, 97, 166, 170, 176, 187
IR (voltage) drop 125
IR compensation 84, 187
Isolation transformer 19, 61, 63, 65, 71, 72, 74, 93, 94, 123, 188

L

Large scale integrated (LSI) 63, 66
Lbft per ampere 111-113, 151-153
Locked rotor 34, 188

M

Magnetic 7, 15, 16, 19, 24, 25, 31, 33, 34, 45, 48, 67, 68, 78, 90, 124
Maintenance 5-9, 15, 19, 25, 121, 164, 166, 170, 171, 173, 177
Mechanical brake 10, 121
Mechanical drives 3, 4
Microprocessor 63, 65, 66, 81

N

Neutral 16, 26, 27, 28, 29, 72, 74

O

OPAMPS 8
Outlet damper 142, 145, 146, 148, 149
Overload 28, 36, 44, 47, 51, 74, 77, 78, 80, 89, 97, 100, 102, 103, 109, 110, 118, 120, 122, 177, 178, 188

P

Parameter 63, 66, 68, 77-93, 96
Payback 148-152, 168, 171
Plugging 11
Poles 15, 16, 28, 31, 33, 45, 47
Power factor 9, 12, 19, 60, 62, 64-66, 68, 70, 74, 95, 139-141, 143, 148, 149, 156, 158, 162, 163, 179, 188
Pressure 3, 5, 132, 136, 144, 145, 147, 148, 150-152, 170, 173, 175
Pull-up torque 35, 36, 188
Pulse width modulated (PWM) 63-65, 93, 95, 141, 143, 188

R

Radiation 19, 124, 129, 131, 134
Ramp-to-stop 79, 84, 85
Ramping 10, 40, 42, 80, 84, 85, 87-91, 175
Reference 19, 20-22, 68, 77, 78, 83, 93, 118, 123, 128
Regeneration 8, 10, 11, 23, 65, 66, 81, 85, 97, 115, 189
Regulation 6, 16, 21, 22, 55, 66, 74, 90, 124, 173, 189
Reverse 17, 23, 28, 69, 72, 79, 87
Ride-through 60, 62
Ripple factor 23-25
Rotating DC drives 7, 8, 19
Rotor 7, 11, 16, 31, 33, 34, 36, 45, 47-50, 55, 66-68, 79, 84, 85, 90, 109
Running 11, 21, 34, 38, 42, 45, 55, 62, 80-83, 97, 102, 110, 111, 118, 142, 166

S

Salient pole 45
Scalar 65, 70
Silicon controlled rectifier (SCR) 8, 9, 19, 23-25, 40-42, 57, 60, 61, 71, 72, 95, 143, 189
Skip frequency 80, 81, 88
Slip 6, 34, 35, 47-49, 51, 53-56, 66, 78, 81, 90, 111-113, 153, 189
Slip compensation 80, 90, 91
Slip rings 45, 48-50, 57
Solid state AC 8, 9, 11, 40-42, 57
Solid state DC 8, 9, 11, 19-22, 57
Speed amplifier 20, 21
Speed regulation 16-19, 21, 22, 33, 34, 42, 44, 47, 49, 50, 55, 80, 90, 113, 124, 165, 179, 175, 189
Squirrel-cage 11, 31, 32, 34, 36, 45, 47, 49, 51, 96, 138
Stator 11, 15, 16, 31, 33, 34, 45, 49, 50, 55, 66, 67, 85, 90
Synchronous (Syn) speed 33-35, 45, 47, 53, 56, 62, 112, 113, 163, 190
Synchronous motor 45, 47, 78, 90, 138

T

Tachometer 6, 22, 66, 122, 125, 190
Temperature 8, 13, 100, 102, 103, 122, 127-129, 131-135, 154, 155, 161, 164, 172, 173, 179
Torque 3-8, 10-12, 15-18, 22, 23, 31, 33-35, 37-40, 42, 45, 47-49, 51, 53-57, 65, 66, 68-70, 72, 78, 79, 81, 84, 97, 102, 107-115, 117-120, 124, 125, 143, 147, 150-153, 163, 166, 177-180, 190
Transformer 11, 19, 25, 33, 61, 63, 65, 71, 72, 74, 93, 94,123, 140, 141
Transistors 8, 57, 63, 68, 71, 78, 79, 81, 91, 159, 190

V

Variable frequency 9, 47, 50, 51, 54, 57, 62, 72, 95-97, 111, 137, 150, 156, 161, 169, 171, 175
Variable speed 1, 2, 3, 4, 9, 50, 57, 96, 118, 121, 147-150, 154, 156, 168, 169, 170, 173, 175, 177
Variable torque 3, 45, 118
Variable Voltage Input (VVI) 57, 60, 143, 190
Variable volts 19, 57, 60-62, 95, 124, 125, 150
Vector 65-70, 115
Volts-per-hertz (V/Hz) 54-57, 60, 63, 78, 84, 86, 92
Vortex 129, 135, 136,

W

Watt losses 13, 128, 132, 135
Wound-rotor motor 47-50, 13